SPRINGER TRACTS IN MODERN PHYSICS

Ergebnisse der exakten Naturwissenschaften

Volume 50

Editor: G. Höhler

Springer-Verlag Berlin Heidelberg GmbH 1969

Manuscripts for publication should be addressed to:

G. Höhler, Institut für Theoretische Kernphysik der Universität, 75 Karlsruhe, Kaiserstraße 12

Proofs and all correspondence concerning papers in the process of publication should be addressed to:

E. A. Niekisch, Kernforschungsanlage Jülich, Institut für Technische Physik, 517 Jülich, Postfach 365

ISBN 978-3-662-15895-1 ISBN 978-3-540-36172-5 (eBook)
DOI 10.1007/978-3-540-36172-5

Originally published by Springer-Verlag Berlin Heidelberg New York in 1969.

Softcover reprint of the hardcover 1st edition 1969

Current Algebra and Phenomenological Lagrange Functions

Invited Papers presented at the
first international Summer School
for Theoretical Physics
University of Karlsruhe
(July 22—August 2, 1968)

Contents

Dynamical Groups and their Currents

A Model for Strong Interactions

A. O. BARUT

Contents

1. Preliminaries

1.1 The Algebraic Description of Particle States and of Currents

The important feature of the dynamical group approach is that it specifies completely from the beginning the particle states as well as the algebra of currents. This aspect is very desirable and leads consequently to more powerful results than either the group theory of multiplets alone, or the algebra of currents alone. Let us, therefore, explain first why this is so.

Group representations in particle physics have been used in the past in three different ways: (1°) to describe the multiplet structure of quantum systems, such as the $O(4)$-classification of the Coulomb-levels, the classification of particle states according to Poincaré group (single particle states), or according to compact groups like $SU(2)$, $SU(3)$, $SU(4)$, $SU(6)$, etc.; (2°) to construct invariant S-matrix elements, like the Poincaré and T, C, P-invariant S-Matrix elements, also like the $SU(2)$-invariant S-matrix elements for strong interactions; (3°) to describe the commutation properties of currents if the interaction Hamiltonian is written as a product of currents, i.e. the interaction currents have among themselves the same commutation relations as those of the generators of some group like $SU(2)$, or $SU(2) \otimes SU(2)$, etc.

Among all the groups the group $SU(2)$ for isospin is common to all the three categories: the group $SU(2)$ specifies the multiplets, the strong interaction S-matrix is invariant under $SU(2)$ [more precisely the group of the S-matrix is the direct product Poincaré $\otimes SU(2)$], and the isospin currents satisfy the commutation relations of $SU(2)$. It was, therefore, tempting to expect the same thing for all other higher groups. Thus when the $SU(3)$, and later, the $SU(6)$-classification of particle states were introduced, one has immediately attempted to construct $SU(3)$- and $SU(6)$-invariant S-matrix elements, and $SU(3)$, $SU(6)$-currents but with no good results. Hence the classification and the invariance of the S-matrix are in general not the same thing. A familiar example is the case of the H-atom: although each leg of the S-matrix in the H-H-scattering, for example, has the $O(4)$-classification, the S-matrix does *not* have the $O(4)$-symmetry at all, it has the symmetry of the Hamiltonian of H_2-molecule.

Other instructive examples are: (1°) the $SU(3)$-classification of strongly interacting particles on the one hand, and the "rotated" (by the Cabibbo angle θ) $SU(3)$ commutation relations of the weak leptonic currents, (2°) the $SU(6)$ classification of particle on the one hand, and the entirely different $SU(6)_W$-invariance of the S-matrix for collinear processes.

The distinction between the group of classification and that of the S-matrix depends, as the above $O(4)$ example shows, whether the forces that give rise to multiplets are the same as the forces between the "particles". Thus, the forces between the atoms are weaker than the Coulomb forces, the forces between the nucleons are weaker (measured by the ratio "Binding energy/reduced Mass") than the forces "inside" the nucleon (forces responsible for the excited states of nucleons). By the same measure the meson-nucleon forces are stronger than the nucleon-nucleon forces and, indeed, the $SU(3)$ and $SU(6)$ classifications work much better for the meson-baryon system, than for the baryon-baryon system.

Thus we have two problems. One relating to the origin of multiplets and of all possible states of the system and the other relating to the properties of the S-matrix or currents. And clearly the situation is more complex than one has thought. The postulates of the current algebra give a partial answer. It is partial, because the evaluation of these commutators is not specified, one has to "saturate" these relations by particle

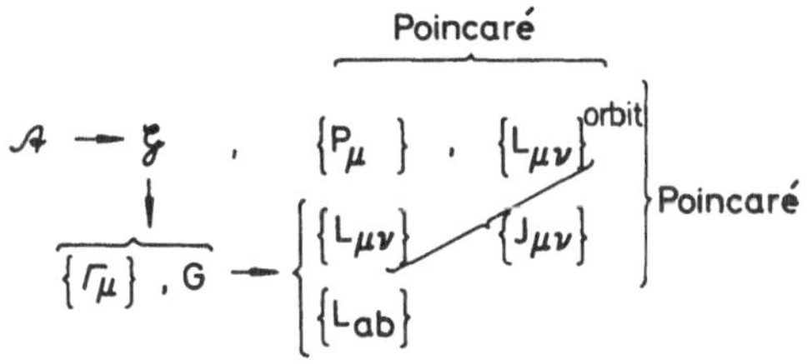

Fig. 1. The complete algebra $\mathscr{A}$ is generated by the dynamical group $\mathscr{G}$ and P_μ (and $L_{\mu\nu}^{\text{orb.}}$); $\mathscr{G}$ in turn contains elements generated by the vector Γ_μ and the group of quantum numbers G, the latter in turn is generated by $L_{\mu\nu}$ (spin) and other degeneracy quantum numbers L_{ab}. Finally Poincaré group is generated by $J_{\mu\nu}$ and P_μ

states whose relation to the algebra of currents is not known. What one then obtains are sum rules relating one amplitude to another.

In the models we are going to discuss the algebraic structure of the theory contains from the beginning (a) a subalgebra G including the homogeneous Lorentz group whose representations give all the rest-frame states of the system, (b) a subalgebra of momentum and currents (Fig. 1).

The particle states at rest are assigned to irreducible representations of G whose subgroup $L_{\mu\nu}$ is identical with the rest frame values of the generators $J_{\mu\nu}$ of the Lorentz group; thus the states are labelled, among other quantum numbers, by spin. The Poincaré group is generated by $J_{\mu\nu}=L_{\mu\nu}+L_{\mu\nu}^{\text{orb.}}$ and by P_μ. The space of particles also admits a vector operator Γ_μ and is actually the representation space of a larger group $\mathscr{G}$. $\mathscr{A}$ is the complete algebraic structure and is, in general, not a Lie algebra but an associative algebra. Now if we know the representation of the complete structure, the matrix elements of the currents (consisting of Γ_μ, P_μ and their combinations), or of other interaction operators, between the particle states are completely specified. Also specified are the spectra of the Casimir operators of the subgroups, such a $P_\mu P^\mu$, the mass operator. This is the basic idea and we are going to show how it is carried out and how the results are interpreted physically.

Clearly the solution outlined here answers on the one hand the mathematical problem of enlarging the Poincaré group with the inclusion of additional quantum numbers such that $P_\mu P^\mu$ has a definite discrete spectrum, and on the other hand, the role of the approximate symmetries in particle physics is understood in terms of an internal structure*.

1.2 Development of the Algebraic Theory of Strongly Interacting Particles

The Algebraic Approach in relativistic particle physics (as distinct from field theory) contains first of all the representation theory of the *Poincaré* group [1]. This theory gives us single particle states and the physical quantities like spin, momentum and mass operators in the form of group generators and Casimir operators. The algebraic approach contains secondly the algebra of the phenomenological internal quantum operators of isospin I and hypercharge Y. The isospin algebra is well established and, as we have seen, is an exact symmetry group of both the multiplets and of the S-matrix of strong interactions. The investigations on the complete algebra of internal quantum numbers culminated in the group $SU(3)$ [2]. This is now a very approximate invariance group of the S-matrix. The symmetry breaking can be taken into account by postulating the commutation relations of the currents only [3]. Another algebraic way was the suggestion to try to generalize the Poincaré algebra such that $P_\mu P^\mu$ has a discrete spectrum [4, 5]. At the same time the concept of "*non-compact dynamical group*" was introduced, a group whose representations give all the rest-frame states of a quantum system having internal degrees of freedom [4, 6]. This concept has been further developped in great detail [7, 8, 9, 10]. Immediately appeared the theorems which state that the generalized framework containing the Poincaré group and the internal quantum numbers with a discrete mass spectrum cannot itself form a Lie algebra [11, 12]. Soon thereafter the group $SU(6)$ was proposed [13] which does, partly, combine the internal quantum numbers and the Lorentz group (in this case just the spin part of the Lorentz group) and which seemed to describe correctly the classification and the static properties of particles. $SU(6)$ has been immediately

* The empirical evidence for a composite structure of hadrons as well as a comparison of the structure and properties of proton with the H-atom has been given in a previous review "Application of the Dynamical Group Theory to the Structure of the Hadrons", in *Lectures in Theoretical Physics,* Vol. X-B (*Gordon* and *Breach*, 1968). These lectures are complementary to and an extension of this earlier review; for a number of mathematical details we refer to this paper. For the relation of the present approach to Regge-pole and high-energy-phenomenology see my contribution in the *Proceedings of Conference on Hadron Spectroscopy*, Acta Phys. Hungaricae, Vol. **26**, No. 1 (1969).

generalized to a formally covariant form of the type $SU(6,6)$ [14]. However, the S-matrix cannot be, on general grounds, simply an index invariant function with respect to such groups, but groups like these can perfectly be taken to be the rest frame groups G (Fig. 1). We also mention at this point the developments in the dynamical groups of strong-coupling models [15]. Meanwhile the problem of including the momentum P_μ into the rest frame dynamical group G have been further investigated, algebraically or by using infinite component wave equations [16]. These investigations received impetus from the group theoretical formulation of the interactions of the H-atom [17]. Subsequently, the relativistic generalization of these theories were applied extensively to the properties of hadrons [18]. Along with these developments the infinite multiplets were required to solve the commutation relations of current algebra [19], and attempts were made to construct local field theories based on infinite component wave equations [20]. But due to the extra space-like solutions of the infinite component relativistic wave equations, it appears that a strict local field theory in the usual sense is not possible [21]. In these lectures we regard the theory as the relativistic quantum theory of composite particles. Finally, the developments using phenomenological Lagrangians show that if the currents transform under the algebra of $SU(3) \otimes SU(3)$ then the multiplets have entirely different group properties; they are associated in these theories with non-linear realizations of the algebra [22]. Recently non-compact group theoretical methods have also been used in connection with the Regge pole expansion of the scattering amplitude [23].

1.3 The O (4,2)-Models

The $O(4,2)$-model of hadrons is a generalization of the Dirac theory on the one hand, and, of the H-atom interactions on the other, both of which can be formulated in $O(4,2)$ in a very similar way. The model takes its relativistic invariance from the first and its compositeness nature from the second. The important feature from a physical point of view is that each hadron is no longer treated as a single elementary particle of mass m and spin j, but that all hadrons with the same internal quantum numbers are considered as a single relativistic system described by a relativistic wave function. The mass spectrum is discrete, and $P_\mu P^\mu = M^2$ is obviously no longer an invariant of the whole theory.

1.4 Notations

Indices μ, ν, ... run from 0, 1, 2, 3; indices i, j, ... run from 1, 2, 3 and indices a, b, ... run from 1, 2, 3, 4, 5 = 0,6. The metric in the six dimensional space is $g_{ab} = - - - - + +$. The metric in the momentum space is $P^2 = P_0^2 - \boldsymbol{P}^2 = m^2$. We use units $c = \hbar = 1$.

A scalar product in the spinor space will be denoted by $(a|b)$, for unitary representations, and by $[a|b]$, for non-unitary representations. The scalar product in the physical Hilbert space is $\langle a|b\rangle$. The one-particle states labelled by momenta and discrete and continuous internal quantum numbers n and λ, respectively, are normalized to

$$\langle n'\lambda' p'|n\lambda p\rangle = \delta_{n'n}(2\pi)\,\delta(\lambda'-\lambda)\,(2\pi)^3\,\delta^{(3)}(\boldsymbol{p}'-\boldsymbol{p})\frac{p_0}{m}. \tag{1.1}$$

We denote the groups by large letters, e.g. $SO(4,2)$, the Lie algebras by small letters, e.g. $so(4,2)$, etc. Group elements will be written as $e^{i\alpha L}$ so that $L^+ = L$, for unitary representations.

2. The General Theory

2.1 The Group O (4,2) and the Physical Interpretation of its Generators

Because the group $SO(4,2)$ is common to all the three systems considered in these lectures we first give quite generally the description of its Lie algebra.

We label the 15 generators of $SO(4,2)$ by the anti-symmetric tensor $L_{ab} = -L_{ba}$

$$L_{ab} = \begin{pmatrix} 0 & L_{12} & L_{13} & L_{14} & L_{15} & L_{16} \\ & 0 & L_{23} & L_{24} & L_{25} & L_{26} \\ & & 0 & L_{34} & L_{35} & L_{36} \\ & & & 0 & L_{45} & L_{46} \\ & & & & 0 & L_{56} \\ & & & & & 0 \end{pmatrix} \tag{2.1}$$

with the commutation relations

$$\begin{aligned}[][L_{ab}, L_{cd}] &= -i(g_{ac}L_{bd} - g_{ad}L_{bc} - g_{bc}L_{ad} + g_{bd}L_{ac}) \\ g_{ab} &= (- - - - + +)\,. \end{aligned} \tag{2.2}$$

The Lie algebra has three Casimir operators

$$\begin{aligned} Q_1 &= L_{ab}L^{ab} \\ Q_2 &= \varepsilon_{abcdef}L^{ab}L^{cd}L^{ef} \\ Q_3 &= L_{ab}L^{bc}L_{cd}L^{d}{}_{a}\,. \end{aligned} \tag{2.3}$$

The basis for an irreducible representation of $O(4,2)$ is uniquely labelled in addition to the eigenvalues of the three Casimir operators by the eigenvalues of five mutually commuting operators in the envelopping algebra of $so(4,2)$. The choice of this set depends on physical applications. We now interprete L_{12}, L_{13}, L_{23} as the spin operators $\boldsymbol{S}$ (or total angular momentum in the rest frame). The generators L_{i4} form a vector with respect to the spin rotation group and will be called the Lenz vector $\boldsymbol{A}$. We interprete $L_{is} = M_i$ as the generators of pure Lorentz transformations. With respect to the Lorentz group $(\boldsymbol{S}, \boldsymbol{M})$ the generators

$$\Gamma_\mu = (L_{56}, \Gamma_{i6}) \tag{2.4}$$

$$\Gamma'_\mu = (L_{45}, -\Gamma_{i4}) \tag{2.5}$$

are four-vectors. The remaining generator $L_{46} = S$ is a scalar with respect to the Lorentz group. We shall denote L_{45} also by $T =$ "tilting" generator, for rcason that will be explained later on.

The parity operator P can be an inner automorphism or an outer one. The representation of the extended group $\{O(4,2), P\}$ can be realized either with or without doubling the basis of the representation of $O(4,2)$. The basis system which is also an eigenstate of parity is denoted by $|n, \pm)$, where n denotes collectively all the quantum numbers of $O(4,2)$.

Note that in the above interpretation of L_{ab} the momenta P_μ are not yet included so that the present use of $O(4,2)$ differs entirely from the conformal group interpretation of the same group in which $SO(4,2)$ also contains momenta.

In particular, we shall be interested in the socalled *most degenerate irreducible representations* of $O(4,2)$ characterized by the vanishing of two of the three Casimir operators. (Only the eigenvalue of $Q_1 \neq 0$.) Another way of characterizing these representations is to prescribe additional algebraic relations between the generators, like the Dirac relation $\{\gamma_\mu, \gamma_\nu\} = 2g_{\mu\nu}$. Among these representations are those which remain also irreducible with respect to the subgroup $SO(4,1)$ and contain the representations of $SO(4)$ only once. In this case it is sufficient to label the states by

$$|(q_1);\ njm\pm); \tag{2.6}$$

because each n [representation of $O(4)$] occurs only once there is no need for other quantum numbers to distinguish different n values. It is possible that more complicated representations of $O(4,2)$ will also be useful, although so far only these most degenerate ones have been used extensively.

The most degenerate representations can be of boson type or fermion type depending whether the lowest spin is 0 or 1/2. Fig. 2 shows the weight diagram of three important representations.

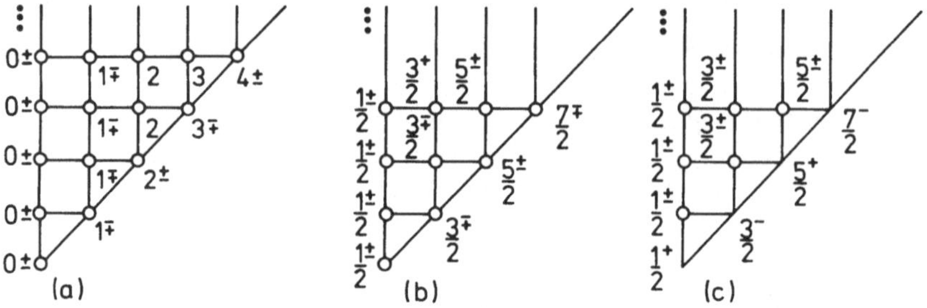

Fig. 2. Range of the quantum numbers n and j shown in weight diagram of the $O(4,2)$ representations. Circles are the doubled states. (a) Boson tower D^B (unitary representation). (b) Fermion tower D^F (unitary representation). (c) Boson tower combined with Dirac matrices (non-unitary representation $n>0$)

2.2 States with Momentum P_μ and the Full Algebra

The representation of $SO(4,2)$ [in general of a group G] defines all the rest frame states that the model contains. Because we also know the generators of pure Lorentz transformations, we can transform the rest frame spinors from $P_\mu=(m, 0, 0, 0)$ to the total momentum

$$P_\mu = m(\mathrm{ch}\,\xi, \hat{\xi}\,\mathrm{sh}\,\xi)\,, \tag{2.7}$$

where $\hat{\xi}$ is the unit vector in the direction of $\boldsymbol{p}$, by the following fundamental equation

$$|n;\,p) = U(\xi)\,|n;\,0) = \mathrm{e}^{i\xi\cdot M}\,|n;\,0)\,. \tag{2.8}$$

In general, P_μ is the total momentum of a many particle state. M is defined in the particular representation of G that we have started with; if $\mathrm{e}^{i\xi\cdot M}$ is not unitary, we write the spinors by $|n;\,p]$. The complete state vectors are

$$|n, p\rangle = |n, p)\,\mathrm{e}^{ipx}\,. \tag{2.9}$$

Thus, by construction

$$P_\mu|n;\,p\rangle = p_\mu|n, p\rangle\,. \tag{2.10}$$

In the general states $|n;\,p)$ the generators $J_{\mu\nu}$ of the Poincaré group also differ from $L_{\mu\nu}$. In x-representation, for example,

$$J_{\mu\nu} = L_{\mu\nu} + \left(x_\mu\frac{\partial}{\partial x_\nu} - x_\nu\frac{\partial}{\partial x_\mu}\right). \tag{2.11}$$

Actually, it seems possible to express completely algebraically P_μ and $J_{\mu\nu}$ in the envelopping algebra of a finitely generated associate algebra [24]. Of course, this will not simply be a Lie algebra containing the

Poincaré group (see Fig. 1). It is also possible to specify the spectrum of $P_\mu P^\mu$ by the assumed conserved currents operators. We discuss this second approach.

2.3 Conserved Currents

We have defined spinors $|n, p)$. But the theory is not yet complete. We have to introduce the quantum mechanical probability density and the probability current (or matter density and matter current). The electromagnetic current may or may not be proportional to the matter current; for a neutral particle like neutron, for example, it is not. Thus

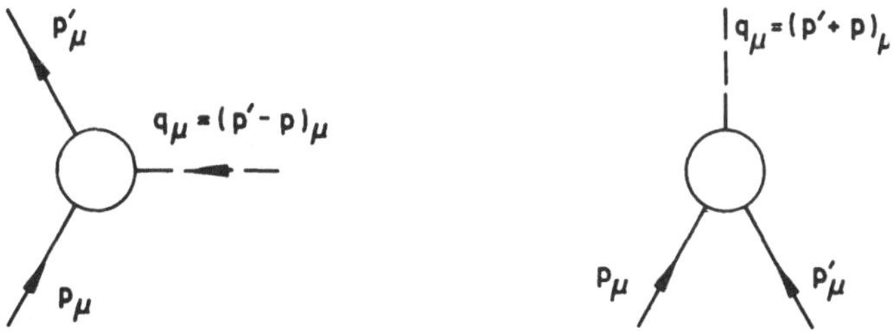

Fig. 3. Scattering and annihilation vertices

we require further the existence of a vector operator which is a function of group generators and momenta

$$j_\mu = j_\mu(L_{ab}, P_\mu, \ldots) \tag{2.12}$$

such that according to general principles of quantum theory the wave functions are normalized

$$(n|j_0|n) = 1\,, \tag{2.13}$$

and the current is conserved:

$$(np|j_\mu(P' \pm P)^\mu|n'p') = 0\,. \tag{2.14}$$

The second equation refers in general to two states of different momenta and different masses as shown in Fig. 3.

Every choice of j_μ leads to a different physical system, because the prescription of a quantum mechanical conserved current operator is essentially equivalent to a specification of the "Hamiltonian" of the system. In order to see this we reduce Eq. (2.15) further into a more tractable form in the case where the theory contains many mass states. Using (2.8) we can write (2.14) in the following form

$$(n|\mathrm{e}^{-i\xi\cdot M} j_\mu P'^\mu \mathrm{e}^{i\xi'\cdot M}|n') = \mp (n|\mathrm{e}^{-i\xi\cdot M} j_\mu P^\mu \mathrm{e}^{i\xi'\cdot M}|n')\,.$$

Because of Lorentz invariance of (2.14) we can choose the rest frame of $P_\mu = m(1,0,0,0)$ and use the relation $e^{-i\xi'\cdot M} j_\mu P'^\mu e^{i\xi'\cdot M} = j_0 m'$ to obtain [25]

$$m_{n'}(n|e^{i\xi'\cdot M} j_0 |n') = \mp m_n (n|j_0 e^{i\xi'\cdot M}|n'). \tag{2.15}$$

Thus unless there are definite mass relations this equation cannot be satisfied. We also see that in the limit $\xi' \to 0$, $|m_{n'}| = |m_n|$, if $(n|j_0|n') \neq 0$, as clearly at zero momentum transfer only degenerate mass states can be reached. We shall discuss the applications of (2.15) in the following sections.

From (2.15) we obtain, taking the + sign and the limit $\xi' \to 0$, the orthogonality of the wave functions for different masses: $(n|j_0|n') = 0$, if $m_n \neq m_{n'}$. This together with (2.13) shows that j_0 is the metric in the spinor space.

We have now defined a complete framework for the states of a system with finite or infinitely many rest frame states. Next we consider the interaction of such a system with external sources which takes the system from one asymptotic state $|n, p)$ to another $|n', p')$. We adopt the S-matrix point of view and write the form of the amplitude for such a process:

$$\text{Amplitude} \propto (np|\,\text{Interaction}\,|n'p'). \tag{2.16}$$

The problem is to specify the interaction. The solution depends of course on the type of interaction and on the type of quantum system. But the hope is that one may eventually discover a universal form of interactions. For the electromagnetic interactions of the system the vertex amplitude (see Fig. 2) is as usual of the form $V_\mu \varepsilon^\mu$, where ε^μ is the polarization vector of the photon and V_μ is the matrix element of an electromagnetic current

$$V_\mu = (n, p|j_\mu^{\text{em}}|n', p'). \tag{2.17}$$

Furthermore, because of the vanishing mass of the photon j_μ^{em} must be conserved:

$$(np|j_\mu^{\text{em}} q^\mu|n'p') = 0, \qquad q_\mu = p'_\mu - p_\mu, \tag{2.18}$$

with

$$(n|j_0^{\text{em}}|n) = \text{charge} = e_n. \tag{2.19}$$

The theory already contains a conserved matter current j_μ. In general, j_μ^{em} is different from j_μ. But because the spinors $|np)$ must satisfy the relations (2.13) and (2.15) there are restrictions on the most general form of j_μ^{em} satisfying (2.18) and (2.19).

Similarly, we can introduce a weak current j_μ^{weak}, and even a strong current, j_μ^{strong}, if the interaction can be written as a product of four-vectors. But, of course, in these cases the interaction operator can à priori be an entirely different operator, a scalar, for example.

3. The O (4,2)-Formulation of the Dirac Theory and its Generalizations

The fundamental irreducible representations of $O(4,2)$ have the dimensions 4, 4 and 6. The two four-dimensional representations are inequivalent. We now take the rest frame states to belong to a four-dimensional irreducible representation given by

$$L_{ab} = \frac{i}{2}\gamma_a\gamma_b\,; \qquad a, b = 1, 2, \ldots, 6\,, \tag{3.1}$$

where

$$\gamma_a = (\gamma_1, \gamma_2, \gamma_3, -\gamma_5, \gamma_0, -i)\,. \tag{3.2}$$

In this special case the representation (3.1) is specified uniquely by the algebraic relation

$$\{L_{\mu 6}, L_{\nu 6}\} = 2g_{\mu\nu}\,; \qquad \mu, \nu = 1, 2, 3, 5\,. \tag{3.3}$$

It is also specified by the eigenvalues of Casimir operators: $q_1 = 15/2$, $q_2 = 0$, $q_3 = 0$.

The subgroup content is now very simple. Because $L^2 = 3/4 = j(j+1)$, only one spin value $j = 1/2$ occurs twice. Two 2-dimensional representations of the subgroup $O(4)$ occur, and because the $O(4)$-Casimir-operators are $L^2 + A^2 = 6/4$ and $L \cdot A = -i\frac{3}{4}\gamma_0$, these two representations can be distinguished by the eigenvalues of γ_0. Also the homogeneous Lorentz subgroup (with Casimir operators $L^2 - M^2 = 0$, $\boldsymbol{L} \cdot \boldsymbol{M} = -i\frac{3}{4}\gamma_5$) is represented twice. A basis of the representation space is then given by the eigenvalues of L_{56}, L^2 and L_{12}: $|njm]$. The operator γ_0 can also be chosen to be the parity operator, it has the required properties, and does not double the Hilbert space (inner automorphism). Thus the parity of the state is equal to the sign of n. The four basis states in the $|njm]$-basis are then

$$|\tfrac{1}{2}, \tfrac{1}{2}, \tfrac{1}{2}, +]\,, \quad |\tfrac{1}{2}, \tfrac{1}{2}, -\tfrac{1}{2}, +]\,, \quad |-\tfrac{1}{2}, \tfrac{1}{2}, \tfrac{1}{2}, -]\,, \quad |-\tfrac{1}{2}, \tfrac{1}{2}, -\tfrac{1}{2}, -]\,. \tag{3.4}$$

Because the 4-dimensional representation is not unitary we have labelled the spinors by the squared brackets $|n]$. The spinors for finite momentum p_μ are now given by

$$\begin{aligned} |n, p] &= e^{\frac{1}{2}\gamma_i\gamma_0\xi^i}|n] \\ &= \left\{\mathrm{ch}\frac{\xi}{2} + \hat{\xi}\begin{pmatrix} 0 & \boldsymbol{\sigma} \\ \boldsymbol{\sigma} & 0 \end{pmatrix}\mathrm{sh}\frac{\xi}{2}\right\}|n]\,, \end{aligned} \tag{3.5}$$

where we have used the γ-matrices with

$$\gamma_i^2 = -1\,, \quad \gamma_0^2 = 1\,, \quad \overset{+}{\gamma_i} = -\gamma_i\,, \quad \overset{+}{\gamma_0} = \gamma_0\,, \tag{3.6}$$

$$\gamma_5 = \gamma_0\gamma_1\gamma_2\gamma_3\,; \quad \gamma_5^2 = -1\,, \quad \overset{+}{\gamma_5} = -\gamma_5\,. \tag{3.7}$$

Next we introduce the probability current. Here à priori there are different possibilities.

Case I. Dirac theory:

$$j_\mu = \gamma_\mu \,. \tag{3.8}$$

The rest frame spinors are now normalized according to

$$[n|\gamma_0|n] = 1 \,. \tag{3.9}$$

In the usual positive metric this equation is equivalent to $(n|\gamma_0\gamma_0|n) = (n|n) = 1$. Similarly the current conservation is expressed as

$$[np|j^\mu q_\mu|n'p] = (np|\gamma_0 j^\mu q_\mu|n'p') = 0 \,. \tag{3.10}$$

From Eq. (2.15) we then obtain immediately

$$m_{n'}n' = \mp m_n n \,. \tag{3.11}$$

This equation with $m_{n'} = m_n$ is satisfied for $n = n'$ only for the + sign and for $n = -n'$ only for the − sign (annihilation) and express, of course, not the charge but the particle-antiparticle conservation (conservation of total n!) because we are dealing with the matter current [26].

The electromagnetic current can now be introduced as a general four-vector in terms of γ_μ and the momenta

$$\begin{aligned} j_\mu^{\text{em}} &= \alpha_1 \Gamma_\mu + \alpha_2 P_\mu + \alpha_3 P_\mu L_{46} + \alpha_4 L_{\mu\nu} q^\nu \\ &\quad + \alpha_2' q_\mu + \alpha_4' L_{\mu\nu} p^\nu + \cdots , \end{aligned} \tag{3.12}$$

$$P_\mu = (p' + p)_\mu \,, \qquad q_\mu = (p' - p)_\mu \,.$$

The electromagnetic vertex amplitude is proportional to the matrix elements of j_μ^{em} between the spinors (3.5) satisfying (3.9) and (3.10). Because of this and because of the fact that the electromagnetic current is also conserved we get restrictions in the form (3.12): the terms with α_2', α_4' can never be conserved for $q^2 \neq 0$, the one with α_4 is separately conserved (the term with α_3 violates, in the present case, parity) and finally there is a relation between the terms with coefficients α_1, α_2 and α_4, namely

$$[np|i\sigma_{\mu\nu} q^\nu|np] = [np|(2m\gamma_\mu + P_\mu)|np] \,, \tag{3.13}$$

because $[np|\gamma^\mu P_\mu|np] = m_n[np|np]$. Consequently, the most general electromagnetic current in this case, assuming parity conservation, is

$$j_\mu^{\text{em}} = \alpha_1 \gamma_\mu + \alpha_4 \sigma_{\mu\nu} q^\nu \,, \tag{3.14}$$

a well-known result, of course, which we derived again in order to illustrate the method to be applied in more general cases. The second term does not contribute to charge and we get

$$e = [n| j_0^{\mathrm{em}} |n] = \alpha_1 [n| \gamma_0 |n] = \alpha_1 \star . \qquad (3.15)$$

The vertex amplitudes of Fig. 2 are now given for the electromagnetic current by

$$V_\mu = [np| j_\mu^{\mathrm{em}} |n'p'] = f_{nn'}(\xi, \xi') \qquad (3.16)$$

with the selection rule (3.11). The expression (3.16) can be evaluated as a function of n and n' in terms of hypergeometric functions (which are, however, for $n = \pm 1/2$ ordinary functions). The crossing symmetry can now be expressed as an analytic continuation in the principal quantum number n. In other words the function $f_{nn'}(\xi, \xi')$ describes for $n = n' = +1/2$ the vertex in the space-like region of the momentum transfer and for $n = 1/2$, $n' = -1/2$ in the annihilation region of the momentum transfer [27].

Case II. We now generalize the matter current (3.8) to include a convective part

$$j_\mu = \gamma_\mu + a P_\mu . \qquad (3.17)$$

The new spinors now satisfy $[n|(\gamma_0 + a P_0)|n] = 1$, and a new conservation equation. Consequently, although the reduction to the form (3.14) of the electromagnetic current holds, one obtains two mass values from the conservation of current (3.17) [26, 28]. One can chose j_μ^{matter} and j_μ^{em} independently. Then there are two possible situations. In one case these two currents are proportional, then there are electromagnetic transitions between the two solutions with masses m and m' [28]. In the second case one can take $j_\mu^{\mathrm{em}} = \gamma_\mu$, i.e. minimal coupling, then the two mass states cannot be coupled by electromagnetic interactions; this is the μ-meson interpretation [29].

This result constitutes one of the ways in which excited states and the form factors, in this case magnetic moments, can be described, namely by introducing convective currents. The other more important way is to use the more complicated infinite dimensional representations of the dynamical group that will be discussed in the following sections.

4. The $O(4,2)$-Formulation of the Interactions of the H-atom

How does a composite system like an H-atom interact with external sources if we want to describe this process purely algebraically? First we have to specify the rest frame dynamical group G whose representation

---

$\star$ Actually, the coefficients α_i are tensor operators with respect to the internal quantum numbers, so that the charge is the matrix element of α_1 with respect to the internal quantum numbers. See Section 5.7.

gives us all the rest frame states of the system and then find the current operator. The discrete spectrum of the H-atom is formally in one-to-one correspondence with the most degenerate unitary irreducible boson representation of the group $O(4,2)$. Of the three Casimir operators given by Eq. (2.3) only Q_1 is different from zero. The basis vectors may be labelled by $|nlm\rangle$, or, in parabolic coordinates, by $|n_1 n_2 m\rangle$. The extension by parity does not double the Hilbert space and the parity of the basis vectors is as it should be $(-1)^l$. The range of the quantum numbers is $n = 1, 2, 3, \ldots$ and for each n, l ranges from 0 to $n-1$. In this representation both the basis vectors and the generators L_{ab} can be expressed in terms of creation and annihilation operators in a simple way [17].

In spite of this one-to-one correspondence, we will now show that the basis vectors $|nlm\rangle$ are *not* the physical states with respect to which the currents have simple linear-form in the Lie algebra and in the momenta. For this purpose we now rewrite quite generally the Schrödinger equation for a potential $V(\boldsymbol{r})$

$$(-\tfrac{1}{2}\Delta + V(\boldsymbol{r}) - E)\tilde{\Psi} = 0\,, \quad (\hbar = m = 1) \tag{4.1}$$

in algebraic form in terms of the group generators L_{ab} alone. There are a number of ways of doing this [17, 30]. Perhaps the simplest one is to realize that the usual angular momentum operators L_i, together with the following operators

$$\begin{aligned} L_{46} - L_{56} &= r\,, \\ L_{46} + L_{56} &= r\Delta\,, \\ L_{i5} - L_{i4} &= X_i\,, \\ L_{i5} + L_{i4} &= -(X_i\Delta + 2X_K\partial^K\partial_i + 2\partial_i)\,, \\ L_{i6} &= -ir\partial_i\,, \\ L_{45} &= i(X_K\partial^K + 1)\,, \end{aligned} \tag{4.2}$$

satisfy the commutation relations of the algebra $so(4,2)$. The differential operators are hermetian if the metric is

$$(\tilde{\Psi}_1, \tilde{\Psi}_2) = \int \tilde{\Psi}_1^*(\boldsymbol{r})\frac{1}{r}\tilde{\Psi}_2(\boldsymbol{r})\,\mathrm{d}^3r\,. \tag{4.3}$$

Consequently, writing in Eq. (4.1)

$$V(\boldsymbol{r}) = \frac{1}{r} f(r, X_i) \tag{4.4}$$

and multiplying the whole equation by r, we obtain

$$[-\tfrac{1}{2}(L_{46} + L_{56}) + f\{(L_{46} - L_{56}), (L_{i5} - L_{i4})\} - E(L_{46} - L_{56})]\tilde{\Psi} = 0$$

or

$$[(E - \tfrac{1}{2})L_{56} - (E + \tfrac{1}{2})L_{46} + f\{(L_{46} - L_{56}), (L_{i5} - L_{i4})\}]\tilde{\Psi} = 0\,. \tag{4.5}$$

Clearly this equation is simplest for a central potential and in the special case of Coulomb potential, where the last term in (4.5) is a function of $(L_{46}-L_{56})$ only, or just a constant, respectively. We remark that

(a) the use of the group $O(4,2)$ is by no means restricted to the special symmetry of the Coulomb problem;

(b) Eq. (4.5) allows now relativistic generalizations which would be very hard to obtain in the Schrödinger formulation, as we shall see.

In order to diagonalize (partly) Eq. (4.5) we put

$$\tilde{\Psi}_{nlm}=e^{i\theta_n L_{45}}\Psi_{nlm} \tag{4.6}$$

and multiply the equation from the left by $e^{-i\theta_n L_{45}}$, to obtain

$$[(E_n-\tfrac{1}{2})(\cosh\theta_n L_{56}+\mathrm{sh}\,\theta_n L_{46})-(E_n+\tfrac{1}{2})(\cosh\theta_n L_{46}+\mathrm{sh}\,\theta_n L_{56}) \\ +e^{-i\theta_n L_{45}} f e^{i\theta_n L_{45}}]\,\Psi_{nlm}=0\,.$$

The coefficient of the L_{46}-term, $(E_n-\frac{1}{2})\,\mathrm{sh}\,\theta_n$, vanishes if $\mathrm{th}\,\theta_n=\dfrac{E_n+\frac{1}{2}}{E_n-\frac{1}{2}}$, or $\theta_n=\frac{1}{2}\ln\left(\dfrac{1}{-2E_n}\right)$. The remaining equation is

$$\{[(E_n-\tfrac{1}{2})\,\mathrm{ch}\,\theta_n-(E_n+\tfrac{1}{2})\,\mathrm{sh}\,\theta_n]\,L_{56}+e^{-i\theta_n L_{45}} f e^{i\theta_n L_{45}}\}\,\Psi_{nlm}=0\,. \tag{4.7}$$

In the special case $f=\text{const.}$, Eq. (4.7) is diagonal in the basis $|nlm\rangle$, where L_{56} has the eigenvalues n, and one obtains the spectrum $E_n=-1/2n^2$ and $\theta_n=\ln(n)$, for $E_n<0$. For $E_n>0$, the solution exists only if $n\,\mathrm{ch}\,\theta_n, n\,\mathrm{sh}\,\theta_n$ are real and if both n and $\mathrm{ch}\,\theta_n, \mathrm{sh}\,\theta_n$ are pure imaginary. In this case we start from a continuous basis $|\lambda lm)$ in which L_{46} is diagonalized with a continuous spectrum λ. This basis is normalized to a δ-function [31]. Now we define the analog of Eq. (4.6), i.e. $|\overline{nlm})=e^{i\theta_n L_{45}}|nlm)$, by

$$|\overline{\lambda lm})=e^{i\theta_n L_{45}}|\lambda lm)\,. \tag{4.8}$$

It is this set of so-called "*tilted states*" $|\overline{nlm})$ and $|\overline{\lambda lm})$ together that correspondends to the set of Schrödinger wave functions for discrete and continuous spectrum, respectively. It follows from this correspondence that the tilted states form a complete set. The basis states $|nlm)$ also form a complete set, the reason that the tilted states $|\overline{nlm})$ alone [without $|\overline{\lambda lm})$] are not complete, is because the transformation (4.6) is not unitary (the tilting parameter θ depends on n!).

The states of the atom of momentum $P_\mu=m(\mathrm{ch}\,\zeta,\ \hat{\zeta}\,\mathrm{sh}\,\zeta)$ are defined according to (2.8) by the spinors

$$|\bar{n};p)=e^{i\zeta\cdot M}|\bar{n})\,, \tag{4.9}$$

where we write simply $|\overline{n})\equiv|\overline{nlm})$, and the M_i are either the generators of the Galilei group or of the Lorentz group depending whether the external motion of the system is treated non-relativistically or relativistically:

$$\begin{aligned} M_i &= L_{i5} - L_{i4} \quad \text{(Galilei)}^{\star}, \\ M_i &= L_{i5} \qquad\qquad \text{(Poincaré)}. \end{aligned} \tag{4.10}$$

Note that in (4.9) the Lorentz transformation has to be applied after the tilting operation to insure the invariance of the theory.

According to our general formalism we have now to introduce the current operator which has a simple form with respect to the tilted states:

$$\begin{aligned} j_\mu &= [(L_{56} - L_{46}), L_{i6}] \quad \text{(Galilei)}, \\ j_\mu &= (L_{56}, L_{i6}) \qquad\qquad \text{(Poincaré)}. \end{aligned} \tag{4.11}$$

We have then the normalization condition,

$$(\overline{n}|j_0|\overline{n}) = (\overline{n}|(L_{56} - L_{46})|\overline{n}) = 1 \tag{4.12}$$

the current conservation condition

$$(\overline{n}p|j^\mu q_\mu|\overline{n}'p') = 0 \tag{4.13}$$

and the explicite expression for form factors

$$V_\mu = (\overline{n}|j_\mu e^{i\xi\cdot M}|\overline{n}). \tag{4.14}$$

These are the basic equations of the theory. We indicate here the following results that follow from these equations:

(1°) The mass spectrum resulting from (4.13) coincides with the previous solution (4.7) and can be written as

$$M_n^2 = M_{\min}^2 + \tfrac{1}{2} m\alpha^2 \frac{n^2-1}{n^2}. \tag{4.15}$$

(2°) The spinors satisfying (4.13) also satisfy the equation describing the motion of the system as a whole.

(3°) When the external motion is treated relativistically, the electromagnetic current can be written in the form

$$j_\mu^{\text{em}} = \alpha_1 \Gamma_\mu + \alpha_2 P_\mu + \alpha_3 P_\mu L_{46}, \tag{4.16}$$

which is of the same type as Eq. (3.12). The mass formula resulting from this current agrees to order α^2 with the mass formula of the H-atom

$$M = M_0 + \lambda \frac{n^2-1}{n^2}. \tag{4.17}$$

$\star$ The relation to the non-relativistic Heisenberg-variables is given by $x_i = (L_{i5} - L_{i4})$, $p_i = (L_{56} - L_{46})^{-1} L_{i6}$.

In this form one can take

$$\alpha_2 \approx 0, \quad \alpha_1 = 1 \quad \text{and} \quad \alpha_3 = \frac{1}{2m_p}. \tag{4.18}$$

The relativistic wave equation is then*

$$(j^\mu P_\mu + \beta L_{46} + \gamma)\tilde{\psi} = 0. \tag{4.19}$$

The transition form factors between arbitrary discrete and continuous states can be calculated from the expression

$$V_\mu = (\overline{n}, p|j_\mu^{\text{em}}|\overline{n}', p')$$

or (4.20)

$$V_\mu = (\overline{n}, p|j_\mu^{\text{em}}|\overline{\lambda}', p'),$$

respectively. In particular it follows that the singularity of the transition form factor coincides with that of the triangular diagram shown in Fig. 4.

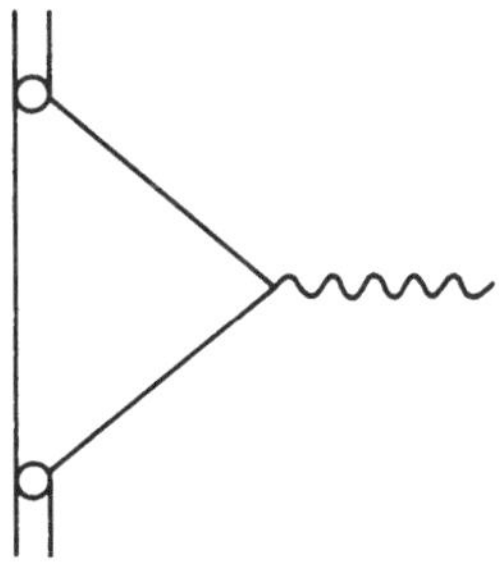

Fig. 4. Diagram whose singularity coincides with that of the form factor calculated group-theoretically

(4°) In order to satisfy the requirement that the charge of all states in the tower is the same, i.e.

$$(\overline{n}|j_0^{\text{em}}|\overline{n}) = e, \tag{4.21}$$

the coefficients α_i in the current must actually be matrices. A simple way to formulate this is to define the physical states with a normalization condition

$$|\text{phys}) = \frac{1}{N_n}|\overline{n}) = \frac{1}{N_n} e^{i\theta_n L_{45}}|n). \tag{4.22}$$

* Actually it turns out that the Eq. (4.19) describes accurately the relativistic effects in H-atom and positronium, including the recoil of the proton (see *Barut, A. O.*, and *A. Baiquni*, Phys. Rev. (in press, 1969)).

In the case of the H-atom it turns out that

$$N_n = n . \tag{4.23}$$

(5°) There are two important lengths ($c = \hbar = 1$) which measure the scale of the composite system and the strength of the interactions:

(a) The Bohr radius $a_0 \sim 1/(m_e \alpha)$ which can be determined by external probes operationally from the slope of the form factor

$$\left.\frac{dG}{dt}\right|_{t=0} \sim \tfrac{1}{2}(m_e \alpha)^{-2} . \tag{4.24}$$

This length is also the range of the "anomalous zitterbewegung" [32] of the composite system when it is treated as a single particle obeying a relativistic wave equation of the type (4.19).

(2°) The second length is the wave-length of the emitted light $a^0/\alpha = (m_e \alpha^2)^{-1}$ which is operationally associated with the slope of the mass-spectrum as a function of the principal quantum number n [see Eq. (4.18)]

$$\left.\frac{dM}{dn}\right|_{n=1} \sim \tfrac{1}{2}(m_e \alpha^2) . \tag{4.25}$$

From these two slopes we can determine α

$$\alpha \sim 2\sqrt{2}\left(\frac{dM}{dn}\right)_{t=1} \cdot \sqrt{\left(\frac{dG}{dt}\right)_{t=0}} . \tag{4.26}$$

We shall apply these operational quantities to the model of proton in the next section.

5. The O (4,2)-Model of Proton

5.1 Basic Hypotheses

We now generalize the ideas of the proceeding sections to a relativistic theory of hadrons. The basic information that we introduce concerns the spectrum of the observed hadron levels. And we want to formulate the assumptions about the spectrum in such a way that the form factors are correctly described.

The simplest possible dynamical group G allowed by a relativistic theory is the Lorentz group $SO(3,1)$ itself. However, if we consider the spectrum and the form factors that can be obtained from $SO(3,1)$ we see that every spin value j can occur only once and that the ground state form factors are of the form $(1 - t/4m^2)^{-3/2}$ [33]. In view of the existence of several states with the same spin and parity (several $1/2^+$-states,

$3/2^{-}$-states) and in view of the empirical form factors of the type $(1-at)^{-2}$ it is necessary to go to a higher group. The choice of $O(4,2)$ is dictated by the remarkable similarities between the properties of baryons and H-atom. The group $O(4,2)$ is appropriate to systems in which one of the constituents is excited with respect to the rest. With this background we now formulate the following two hypotheses:

Hypothesis I. The rest frame states of baryons with the same internal quantum numbers belong to an irreducible representation of the group $O(4,2)$.

Hypothesis II. The conserved current is the most general minimal linear current in the Lie algebra of $SO(4,2)$ acting between the tilted physical states of the type (4.22).

Remarks concerning theses hypotheses: (a) The infinite dimensional representations of $O(4,2)$ introduce infinitely many excited states, few of them actually observed at the present time. In fact, there are in the model infinitely many states with spin parity $1/2^{\pm}$ with increasing mass, infinitely many $3/2^{\pm}$-states, infinitely many $5/2^{\pm}$-states, etc. The introduction of so many hadrons might be objectionable if we would treat each hadron state as an elementary particle and introduce a separate field for each state. However, from the point of view of composite systems one cannot just use only those states that have been observed*. We are reminded that at the time Balmer set up his formula only four levels of the H-atom were known.

(b) The linear current is essentially the one that occurs in the H-atom. Only linear tensor operators are taken for simplicity. For example, the spin-orbit coupling terms are not included in the H-current. It is possible that the correct relativistic version of the H-atom also contains such spin-orbit terms even with linear currents. At any rate the precise form of the current will depend on further properties of the excited states. The dependence on internal quantum numbers will be discussed in Section 5.7.

5.2 Matter Current and Mass Spectrum

The matter current that we assume is

$$j_\mu = \alpha_1 \Gamma_\mu + \alpha_2 P_\mu + \alpha_3 P_\mu L_{46} \tag{5.1}$$

acting on the physical states

$$|\overline{n}) = \frac{1}{N_n} \mathrm{e}^{i\theta_n L_{45}} |njm \pm) . \tag{5.2}$$

* In fact a number of the predicted states have been observed already, see *A. O. Barut*, Physics Letters, **26** B, 308 (1968).

We may use a most degenerate fermion representation with the spectrum $j^P = 1/2^\pm, 3/2^\pm, \ldots (n-1)^\pm$, for each $n = 3/2, 5/2, 7/2 \ldots$.

The mass formula that the current (5.1) gives via the Eq. (2.15) is the following [18, 25]

$$M_n^2 = \left[2\left(\alpha_3^2 + \frac{\alpha_2^2}{n^2}\right)\right]^{-1} \left\{\alpha_1^2 + 2\beta\alpha_3 + \frac{2\gamma\alpha_2}{n^2} + \left[\left(\alpha_1^2 + 2\beta\alpha_3 + \frac{2\gamma\alpha_2}{n^2}\right)^2 - 4\left(\beta^2 + \frac{\gamma^2}{n^2}\right)\left(\alpha_3^2 + \frac{\alpha_2^2}{n^2}\right)\right]^{1/2}\right\}, \tag{5.3}$$

where β and α are two new constants. There is a second set of mass values with the $-$ sign in front of the square root. But these solutions do not satisfy the normalization condition

$$\frac{1}{|N_n|^2} (n| e^{-i\theta_n L_{45}} j_0 e^{i\theta_n L_{45}} |n) = 1 \tag{5.4}$$

or

$$|N_n|^2 = (\alpha_1 n \operatorname{ch}\theta_n + 2\alpha_2 m_n + 2\alpha_3 m_n n \operatorname{sh}\theta_n)^{-1}. \tag{5.4'}$$

The tilting angle θ_n in this equation is also determined from the current conservation to be

$$\operatorname{sh}\theta_n = n \frac{\beta - \alpha_3 M_n^2}{\gamma - \alpha_2 M_n^2}. \tag{5.5}$$

The results (5.3)—(5.5) may equivalently be obtained from the wave equation (4.19).

5.3 The Electromagnetic Current

The electromagnetic current may contain the same number of terms as the matter current plus the terms which are automatically conserved:

$$j_\mu^{\mathrm{em}} = a_1 \Gamma_\mu + a_2 P_\mu + a_3 P_\mu L_{46} + i a_4 L_{\mu\nu} q^\nu. \tag{5.6}$$

The elastic and inelastic electromagnetic form factors can now be calculated with this current between any two arbitrary states. An interesting result is the double pole behavior of the magnetic form factor of the ground state

$$G_M(t) = \frac{\mu}{\left(1 - \operatorname{ch}^2\theta \frac{t}{4m^2}\right)^2} \tag{5.7}$$

with

$$\mu = -\tfrac{1}{2} a_1 \operatorname{ch}\theta_1 - m a_4.$$

Here m is the mass of the ground state. The electric form factor $G_E(t)$ has a similar form with some small deviations for large t from the double pole behavior [18a].

A noteworthy point is the position of the singularity of the form factor at $t=t_1=4m^2/\mathrm{ch}^2\theta$. The factor $\mathrm{ch}^2\theta$ is responsible in the case of H-atom for the anomalous singularity. Here also the singularity is displaced from its "normal" value $4m^2$*. Thus, an interpretation suggests itself that the proton form factor can also be described effectively by the triangular diagram of Fig. 3. However, what the constituents are in this case, we do not know.

The form factors in the time-like region, $t>0$, cannot yet be discussed adequately, because the representation that we are using does not contain antiparticles. Although the form factor (5.7) itself is crossing symmetric, it is not the only term that contributes in the annihilation channel: In the scattering channel we have considered a single system and the transitions in that system due to external fields. In the annihilation case, we have to take *two* composite systems that interact. Eq. (5.7) will give a contribution, but there will also be contributions from pionic intermediate states that are responsible for the branch points of the form factor for $t>0$.

Electromagnetic transition form factors to high spin excited states are at the present time being measured and theoretically calculated [34].

We note that Eqs. (5.1) and (5.6) imply an approximate equality of matter and charge distributions in this proton model, approximate because although the last term in (5.6) does not contribute to the charge, it contributes to the electric form factor $G_E(t)$ for $t \neq 0$.

5.4 The Scalar Vertex

The reactions of the type $B' \to B + \pi$, where B and B' are arbitrary baryon states, can be described by the amplitude

$$A = g(\bar{n}p|\pi|\bar{n}'p') \tag{5.8}$$

where π is the simplest linear pseudoscalar operator in $O(4,2)$. There is a single overall coupling constant for all the states in the same tower. Eq. (5.8) actually defines a "*strong form factor*" $G^s(t)$ by equating (5.8) to

$$A = G^s(t)\bar{u}(p)\gamma_5 u(p), \tag{5.9}$$

for the spin 1/2-states, for example. Such strong form factors have not been used so far in strong interaction theory, because their empirical

* Actually due to the 2π-intermediate state in the time-like region there is a branch point in $G(t)$ at $t=4m_\pi^2$. The word "normal" here refers to an intermediate state of mass $2m$.

separation is not easily possible. However, we will later show that they are quite useful in the understanding of diffraction phenomena. The functional form of $G^s(t)$ is quite similar to the electromagnetic form factors, a fact which also will be important in the diffraction scattering. The decay amplitudes are on mass-shell values of the strong form factors and the decay rates will be calculated by

$$\Gamma = |A|^2 \times \text{invariant phase space}\,. \tag{5.10}$$

The decays involving many pions, $B' \to B + \pi + \pi + \cdots$ are more complicated, unless the process goes via a cascade process, i.e.

$$B'' \to B' + \pi \to (B + \pi) + \pi\,, \qquad \text{etc.}$$

For detailed numerical results, which, by the way, agree with the observed decay rates, we refer to the literature [18].

5.5 Diffraction Scattering

The diffraction scattering is defined as that part of the scattering amplitude of two hadrons in which there is no exchange of quantum numbers in the crossed channel. In this case the hadrons only transfer momentum to each other. Thus the simplest form of such an amplitude in the present formalism is

$$\begin{aligned} A = \sum_i g_i [(\overline{n}_3 p_3|O^i|\overline{n}_1 p_1)\,(\overline{n}_4 p_4|O^i|\overline{n}_2 p_2) \\ - \text{Exchange term in the case of identical particles}]\,, \end{aligned} \tag{5.11}$$

i.e. a current-current interaction. Here O^i can be taken to be the scalar, vector, ... operators in $O(4,2)$. The underlying idea is that if the structure is taken into account via the $O(4,2)$-wave function, the interaction may be quite simple. Indeed, the Eq. (5.11) results even in the simplext case of scalar interaction, $O^i = I$, in a non-trivial cross-section formula [35]

$$\begin{aligned} \left(\frac{d\sigma}{d\Omega}\right)_{\text{unpol.}} = \frac{g_s^2}{s}\left[\frac{(1-t/4m^2)^2}{(1-at)^8} + \frac{(1-u/4m^2)^2}{(1-au)^8} - \frac{s/4m^2 - ut/(4m^2)^2}{(1-at)^4(1-au)^4}\right] \\ a = \text{ch}^2\theta/4m^2 \end{aligned} \tag{5.12}$$

for the elastic scattering of two identical particles like p – p. The simple theory (5.12) fits the 90°-proton-proton scattering data extremely well. It is also possible to fit the normalized high energy, high momentum transfer data, $d\sigma/dt/(d\sigma/dt)\,t_0$, at all energies and angles by an interaction of the type $O = e^{i\Delta L_6}$, the most general scalar in $O(4,2)$. The $t = 0$ values cannot yet be evaluated because there are other exchange effects than a simple momentum transfer.

The result (5.12) can be interpreted essentially as the square of the product of two "strong form factors" (Fig. 5).

Thus, aside from exchange terms, we have

$$d\sigma/d\Omega \sim |G^s|^4 .$$

We have emphasized in the previous section that $G^s(t)$ is very similar in functional form (not strength!) to $G_E(t)$. Hence, this provides an explanation for the observed qualitative relation between the p – p-scattering and the electromagnetic form factors, two entirely unrelated phenomena, one strong the other electromagnetic.

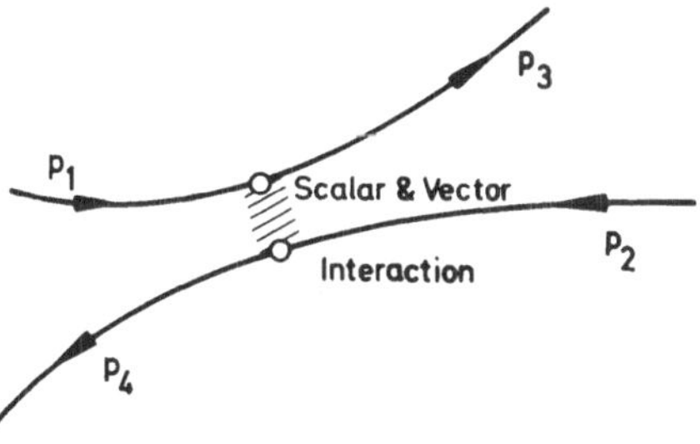

Fig. 5. Kinematics of the elastic scattering and the direct "hard sphere" interaction

We also remark that the present formalism allows one automatically to evaluate arbitrary inelastic diffraction process of the type

$$N + N' \rightarrow N'' + N''' ,$$

once the form of the operators O^i in (5.11) is specified, and, that the resultant amplitude is an analytic function of the spin and other quantum numbers of the external legs of the S-matrix.

5.6 Model for the Proton

The theory and the calculations summarized in this chapter suggest the following composite model of proton: with respect to external electromagnetic interaction it consists of two parts, each interacting (locally) with the external fields as in an atom. It is possible to use the qualitative concepts of atomic physics when analysing the behavior of proton: thus, we can talk of excitation and decay of one part of the proton by external pionic or electromagnetic field, capture of pions in "Bohr orbits", exchange of external pions with the pions in the system, and so on. What the actual constituents are, needs not yet to be precisely specified. However, operationally we can now evaluate the relevant lengths and coupling constants as in Eqs. (4.24)—(4.26).

$$\left(\frac{\mathrm{d}G}{\mathrm{d}t}\right)_{t=0} \sim 2\frac{\mathrm{ch}^2\theta}{4M^2} \cong 2.8\,(\mathrm{GeV}/c)^{-2}\,, \tag{5.13}$$

which is also the characteristic quantity for the zitterbewegung of proton when it is described by an infinite component wave equation [32]. From the mass spectrum (5.3) we obtain further

$$\left(\frac{\mathrm{d}M}{\mathrm{d}n}\right)_{n=1} \sim \tfrac{1}{2}\,(\mathrm{GeV}/c)\,, \tag{5.14}$$

so that Eq. (4.26) gives for the analog of α

$$\text{“}\alpha\text{”} \sim 2\,. \tag{5.15}$$

It is interesting that the length in (5.13) also manifests itself in the diffraction scattering. The first term in the formula (5.12) for small t and at fixed s can be written as

$$\frac{(1-t/4m^2)^2}{(1-at)^8} \cong \mathrm{e}^{\gamma t}\,. \tag{5.16}$$

Expanding this for small t to order t^2 we find

$$\gamma = 8a - \frac{1}{2m^2} \cong 10.7\,(\mathrm{GeV}/c)^{-2}\,, \tag{5.17}$$

which is very close to the magnitude of the observed slope of the diffraction peak [ca. $9\,(\mathrm{GeV}/c)^{-2}$].

5.7 The Inclusion of Internal Quantum Numbers

So far we have considered a single $O(4,2)$ tower with fixed internal quantum numbers. In order to treat decays of the type $\Lambda \to N\pi$, or $\Delta^{(I=3/2)} \to N\pi$, or the electromagnetic transitions of the form $\Delta \to N\gamma$, ... it is necessary to include into the formulation the algebra of internal quantum numbers. One might think immediately to generalize the rest frame group $O(4,2)$ to more general groups like $SU(6,6)$ of $SL(6,6)$ as has been proposed, However, this approach has two main difficulties:

(1°) the form factors given by large groups like $SU(6,6)$ decrease two fast, like $1/(1-at)^6$, and cannot be applied to hadrons,

(2°) these groups predict a large number of states of very high isospin and hypercharge values.

As there is no evidence for such large groups, we include the group of internal quantum numbers, in the rest frame, simply as a direct product to $O(4,2)$; that is we assume

$$G = O(4,2) \otimes SU(3)\,. \tag{5.18}$$

This does not mean that $SU(3)$ is an exact symmetry, but on the contrary, that $SU(3)$ must be necessarily broken. The reason is the following. When we define from the rest frame states

$$|\bar{n}jm, II_3Y).$$

the states with momentum p_μ by

$$|\bar{n}, \alpha, p) = |\bar{n}jm, II_3Y, p) = e^{i\xi_\cdot M}|\bar{n}jm, II_3y), \tag{5.19}$$

the spinors depend only on $\xi = \mathrm{th}^{-1} p/E$. Consequently, the vertex amplitudes

$$(\bar{n}, \alpha, p|O^i|\bar{n}', \alpha', p') \tag{5.20}$$

are universal functions of ξ and of the quantum numbers. But because ξ is different for each member of the multiplet, we will get on the mass shell different values and the symmetry is broken*. Thus, there is a definite prescription how to take the mass differences into account in applying the group $SU(3)$ to reactions, a point which was beset with some ambiguities in the past [36]. It is in this spirit that the decays of baryons and mesons have been calculated [18].

Now to describe the mass differences within multiplets we have to assume that the coefficients in the current

$$j_\mu^{(i)} = \alpha_1^{(i)}\Gamma_\mu + \alpha_2^{(i)}P_\mu + \alpha_3^{(i)}P_\mu S \tag{5.21}$$

and in the wave equation

$$[j_\mu^{(i)}P_\mu + \beta^{(i)}L_{46} + \gamma^{(i)}]\tilde{\Psi}(p) = 0 \tag{5.22}$$

are tensor operators with respect to internal quantum numbers which we have indicated by a superscript i. What kind of tensor operators they are can be inferred by investigating mass formulas and other amplitudes. The electromagnetic current will have also tensor coefficients

$$\overset{(i)}{j_\mu^{\mathrm{em}}} = a_1^{(i)}\Gamma_\mu + a_2^{(i)}P_\mu + a_3^{(i)}P_\mu L_{46} + ia_4^{(i)}L_{\mu\nu}q^\nu. \tag{5.23}$$

The special form of the tensor operators of the coefficients may also be inferred by theoretical assumption. A particular case that satisfies the current algebra charge commutation relation

$$[Q^i, j_\mu^i] = if^{ijk}j_\mu^k$$

at infinite momentum has been given recently [37]. If α's are proportional to the $SU(3)$-generators λ^i, then the current algebra is automatically satisfied, if the spinor states are complete with normalization (5.4).

As an example of application of (5.20)—(5.23) we consider in the next section the weak decays.

* Another origin of symmetry breaking lies in the tilting operation (4.22), because the tilting angle is different for each member of the multiplet.

5.8 The Weak Vertex

Having introduced a structure into the hadrons, the probing of this structure by external electromagnetic and weak interactions is described simply by the vertex amplitude. The electromagnetic vertex has been discussed. For the weak vertex, we have both the non-leptonic processes, like $\Lambda \to p\pi^{\cdot}$, and the semi-leptonic processes, like $\Lambda \to p l \bar{\nu}$. For the non-leptonic case the most direct approach is to use the scalar and pseudoscalar operators in $O(4,2)$ so that the amplitude will be given by

$$A = \left(n, \alpha, p \left| \begin{matrix} s \\ p \end{matrix} \right| n', \alpha', p' \right) \tag{5.24}$$

where s and p stand for scalar or pseudoscalar operators (or for s- and p-wave baryon decay amplitudes). The tensorial property with respect to the internal quantum number α has to be introduced as a further assumption. The new feature of Eq. (5.24) is that it allows immediately to calculate the mass and form factor-correction to an assumed $SU(3)$-property in the rest frame. Interestingly these corrections can be as large as 20% in the decay rate [38], so that indeed more accurate measurements are necessary to discover the true symmetry principles of the non-leptonic decays.

In the leptonic decays, it is natural to generalize the Gell-Mann-Feynman conserved current hypothesis for the vector part of the weak current to $O(4,2)$ [37] so that matter current, electromagnetic current and weak vector current will be all proportional to each other. The axial vector current is not conserved, but we can again relate its divergence to the pseudoscalar vertex so that the ideas of PCAC are generalizable to $O(4,2)$-currents. The calculation of the form factors in the decay $K \to \pi + l + \nu$ and the symmetry breaking effects has been given recently and the results agree with experiments [39].

6. Conclusions

We have presented a discussion of the differences between the group structure of the multiplets and the group structure of interactions. We have also presented models in which these properties are explicitely exhibited. Because both the particle states and the currents are explicitely given not only sum rules but all matrix elements themselves are known. We have reviewed the general framework of the dynamical group theory, its special form in the case of the Dirac particle and the H-atom and its application to hadrons. The rest frame group $O(4,2)$ emerges as the common and the most important ingredient in all three cases. Within this framework, the rest frame group, the particular representation of it, and the form of the quantum mechanical conserved current specify a particular physical system. We have also discussed a *quantitative* model for the hadrons, that describe rather well (a) the mass spectrum, (b) the partial decay rates, (c) the magnetic moments, (d) the electromagnetic form factors and transition form factors, and (e) the diffraction scattering of hadrons.

There are problems which need further study. Among these we have mentioned (1°) the inclusion of antiparticles and the form-factors in the time-like region, (2°) a more general treatment of the scattering processes and of general crossing properties, (3°) a deeper understanding of the tensorial properties of interactions with respect to the internal quantum numbers, and (4°) a more fundamental basis for the several parameters which we have fitted so far using empirical information.

References

1. *Wigner, E. P.:* Ann. Math. **40**, 149 (1939).
2. *Gell-Mann, M.*, and *Y. Ne'eman:* The eightfold-way. New York: Benjamin 1964.
3. — Phys. Rev. **125**, 1067 (1962); — *Ne'eman, Y.:* Algebraic theory of particle physics, New York: Benjamin 1967; — *Renner, B.:* Current algebras and their applications. Oxford: N.Y.: 1968; — *Adler, S. L.*, and *R. F. Dashen:* Current algebras and applications to particle physics. New York: Benjamin 1968.
4. *Barut, A. O.:* Proceedings of the first coral gables conference. San Francisco-London: W. H. Freeman 1964.
5. *Kurşunoğlu, B.:* Proceedings of the first coral gables conference. San Francisco-London: W. H. Freeman 1964.
6. *Barut, A. O.:* Phys. Rev. **135**, B 839 (1964).
7. —, and *A. Böhm:* Phys. Rev. **139**, 1107 (1965).

7a. — *P. Budini*, and *C. Fronsdal:* Proc. Roy. Soc. A **291**, 106 (1966).

8. *Mukunda, N., L. O'Raifertaigh*, and *E. C. G. Sudarshan:* Phys. Rev. Letters **19**, 322 (1965).
9. *Dothan, Y., M. Gell-Mann*, and *Y. Ne'eman:* Phys. Rev. Letters **17**, 148 (1965).
10. See the reviews in: High energy physics and elementary particles (IAEA, Vienna 1965), in: Non-compact groups in particle physics. New York: Benjamin 1966; — Symmetry principles and particle physics. San Francisco-London: W. H. Freeman 1967.
11. *McGlinn, D.:* Phys. Rev. Letters **12**, 467 (1964). — *O'Raifertaigh, L.:* Phys. Rev. Letters **14**, 333 (1965). — *Greenberg, O. W.:* Phys. Rev. **135**, B 1447 (1965). — *Michel, L.:* Phys. Rev. **137**, B 405 (1965). — *O'Raifertaigh, L.:* Phys. Rev. **139**, 1052 (1965).
12. See the review by *Hegerfeldt, G. C.*, and *J. Henning:* Fortschr. Physik **16**, 491—543 (1968).
13. *Gürsey, F.*, and *L. Radicati:* Phys. Rev. Letters **13**, 173 (1964). — See also the book by: *Dyson, F. J.:* Symmetry groups in nuclear and particle physics. New York: Benjamin 1965 for further references.
14. See the review *Delbourgo, R., M. A. Rashid, Abdus Salam*, and *J. Strathdee:* In: High energy physics and elementary particles. Intern. Atomic. Energy Agency, Vienna 1965.
15. *Cook, T., C. Goebel*, and *B. Sakita:* Phys. Rev. Letters **15**, 35 (1965). See the review by *Goebel, C.:* Lectures in theoretical physics, Vol. IX B. New York: Gordon and Breach 1967.

16. *Budini, P.*, and *C. Fronsdal:* Phys. Rev. Letters **14**, 968 (1965). — *Barut, A. O.:* Phys. Rev. **156**, 1538 (1967). —, and *H. Kleinert:* Phys. Rev. **156**, 1541 (1967); **160**, 1149 (1967). — *Nambu, Y.:* Prog. Theor. Phys. Suppl. **37**, **38**, 368 (1966). — *Böhm, A.:* Lectures in theoretical physics, Vol IX B. New York: Gordon and Breach 1967.
17. *Barut, A. O.*, and *H. Kleinert:* Phys. Rev. **156**, 1541 (1967); **157**, 1180 (1967); **160**, 1149 (1967). — *Fronsdal, C.:* Phys. Rev. **156**, 1665 (1967). — *Kleinert, H.:* Lectures in theor. physics, Vol. X B. New York: Gordon and Breach 1968. — *Nambu, Y.:* Phys. Rev. **160**, 1171 (1967).
18. —, *D. Corrigan*, and *H. Kleinert:* Phys. Rev. Letters **20**, 167 (1968). —, and *H. Kleinert:* Phys. Rev. Letters **18**, 754 (1967). — *Kleinert, H.:* Phys. Rev. Letters **18**, 1027 (1967). — *K. C. Tripathy:* Phys. Rev. Letters **19**, 918 (1967); **19**, 1081 (1967). —: Nucl. Phys. B **4**, 455 (1968).
19. *Fubini, S.:* Proc. 4th Coral Gables Conference. San Francisco-London: W. H. Freeman 1967. — *Gell-Mann, M., D. Horn*, and *J. Weyers:* Proc. Heidelberg Conference. Amsterdam: North Holland 1968. — *Leutwyler, H.:* Phys. Rev. Letters **20**, 561 (1968). — *Hamprecht, B.*, and *H. Kleinert:* University of Colorado preprint 1968.
20. *Fronsdal, C.:* Phys. Rev. **156**, 1653 (1967). — *Nambu, Y.:* Proc. of the 1967 Intern. Conference on Particles and Fields. New York: Interscience 1967.
21. *Grodsky, I. T.*, and *R. F. Streater:* Phys. Rev. Letters **20**, 695 (1968).
22. *Gürsey, F.:* Nuovo Cimento **16**, 705 (1960). — *Weinberg, S.:* Phys. Rev. Letters **18**, 507 (1967). — *Wess, J.*, and *B. Zumino:* Phys. Rev. **163**, 1727 (1967). — *Schwinger, J.:* Phys. Letters **24** B, 473 (1967).
23. *Friedman, D. Z.*, and *J. M. Wang:* Phys. Rev. **153**, 1596 (1967). — *Domokos, G.*, and *P. Suranyi:* Nuovo Cimento **56** A, 44 (1968). —, and *G. L. Tindle:* Phys. Rev. **165** 1906 (1968). — *Cosenza, G., A. Sciarrino*, and *M. Toller:* Nuovo Cimento **57** A, 283 (1968). — *Delbourgo, R., Abdus Salam*, and *J. Strathee:* Phys. Rev. **172**, 1727 (1968).
24. *Böhm, A.:* To be published.
25. *Barut, A. O., D. Corrigan*, and *H. Kleinert:* Phys. Rev. **167**, 1527 (1968).
26. — Phys. Rev. Letters **20**, 893 (1968).
27. —, *P. Cordero*, and *G. C. Ghirardi:* IC/68/96 (to be published).
28. — — — Phys. Rev. (in press).
29. — IC/68/59.
30. *Nambu, Y.:* Proc. of the 1967 Intern. Conference on Particles and Fields. New York: Interscience 1967.
31. *Barut, A. O.*, and *C. Phillips:* Commun. Math. Phys. **8**, 52 (1968).
32. —, and *S. Malin:* Rev. Mod. Phys. **40**, 632 (1968). — *Kleinert, H.*, and *S. Malin:* Nuovo Cimento, 58 A, 835 (1968).
33. —, and *H. Kleinert:* Phys. Rev. **156**, 1546 (1967).
34. *Corrigan, D.:* Thesis University of Colorado 1968.
35. *Barut, A. O.*, and *D. Corrigan:* Phys. Rev. **172**, 1593 (1968).
36. *Levinson, C. A., H. J. Lipkin*, and *S. Meshkov:* Phys. Letters **1**, 44 (1962). — Phys. Rev. Letters **10**, 361 (1962). — Review by *Harari, H.*, in: High energy physics (Vienna, IAEA, 1965).
37. *Corrigan, D.*, and *B. Hamprecht:* Lectures in theoretical physics, Vol. XI-A. New York: Gordon and Breach 1969.
38. *Barut, A. O.*, and *S. Malin:* Nuclear Phys. B **9**, 194 (1969).
39. —, and *K. C. Tripathy:* Phys. Rev. (in press).

Prof. Dr. *A. O. Barut*
University of Colorado
Boulder, Colorado/U.S.A.

Current Algebra and Lightlike Charges

H. LEUTWYLER

Contents

1. Introduction

Current operators have come to play a prominent role in particle physics. They occur in two quite different contexts. On the one hand electromagnetic as well as weak interactions are described in terms of current operators. In this context the currents may be regarded as *observables* in the sense that particular matrix elements of certain current components are intimately related to measurable quantities like decay amplitudes and scattering cross sections. On the other hand, currents also arise in connection with *symmetry groups*. In fact, Noether's theorem asserts that if the Lagrangian is invariant with respect to a symmetry group generated by a set of operators $I_1, I_2, \ldots$ then there exists a corresponding set of conserved currents $j_1^\mu(x), j_2^\mu(x), \ldots$ such that the generators are given by

$$I_i = \int d^3x\, j_i^0(\boldsymbol{x}, t) \qquad i = 1, 2, \ldots \tag{1.1}$$

The structure relations

$$[I_i, I_k] = iC_{ikl} I_l \tag{1.2}$$

characteristic of the symmetry group impose important conditions on the conserved currents associated with this symmetry.

Even if the Lagrangian is not invariant with respect to a given set of transformations of the basic fields, Noether's theorem still allows us to construct a set of local currents associated with the generators of the transformation group. In this case the currents are of course not conserved and the generators are not independent of time. Nevertheless algebraic relations of the type (1.2) which characterize the structure of the transformation group rather than the dynamical properties of the system, may survive.

A well-known example of a current which represents an observable and at the same time gives rise to a charge that generates an exact symmetry group is the electromagnetic current. A second example is provided by the vector part of the strangeness conserving hadronic weak current. According to the conserved vector current hypothesis [1] the charge associated with this object is intimately related to the generators of the isospin group. In this case the symmetry group is still an exact one as far as strong interactions are concerned. Current algebra is an attempt to extend this relation between observables and elements with simple algebraic properties to the more general case of a broken symmetry group. We briefly review the basic assumptions involved in current algebra in the next section [2].

2. Charge Algebra

Let us denote by $j_i^\mu(x)$, $(i=1, \ldots, r)$, the basic set of currents constituting the current algebra in question. In the case of the algebra $SU(2) \times SU(2)$ e.g. this set contains a triplet of vector currents as well as a triplet of axial vector currents, in the case of $SU(3)$ the set consists of an octet of vector currents etc. Only a subset of these currents viz. those that are associated with isospin and hypercharge are conserved. Consequently only some of the charges

$$I_i(t) = \int d^3x\, j_i^0(\boldsymbol{x}, t) \tag{2.1}$$

are independent of time. The remainder does not commute with the Hamiltonian and does, therefore, not generate a symmetry group. The basic assumption involved in current algebra is that at equal times the operators nevertheless generate a Lie algebra

$$[I_i(t), I_k(t)] = iC_{ikl} I_l(t)\,. \tag{2.2}$$

This assumption may be sharpened by postulating that the currents transform according to the regular representation under the algebra generated by the operators $I_i(t)$

$$[I_i(t), j_k^\mu(x)] = iC_{ikl} j_l^\mu(x)\,; \qquad (x^0 = t) \tag{2.3}$$

or even that the currents satisfy local commutation rules of the type

$$[j_i^0(x), j_k^0(y)] = iC_{ikl} j_l^0(x)\,\delta(\boldsymbol{x}-\boldsymbol{y})\,; \qquad (x^0 = y^0)\,. \tag{2.4}$$

In the following we restrict ourselves to the algebra generated by the charges $I_i(t)$. We will return to the stronger requirement (2.3) later on. No use will be made of the local commutation rules (2.4).

In order to convert algebraic relations of the type (2.2) into statements about measurable quantities one considers one-particle matrix elements and inserts a complete set of states to obtain a *sum rule* of the form

$$\sum_n \langle p', \alpha' | I_i(t) | n \rangle \langle n | I_k(t) | p, \alpha \rangle - (i \leftrightarrow k) = iC_{ikl} \langle p', \alpha' | I_l(t) | p, \alpha \rangle\,. \tag{2.5}$$

Actually this equation contains a whole family of sum rules since the momentum $\boldsymbol{p}$ of the initial state may be chosen at will. (The final momentum is then fixed by momentum conservation.) This freedom is intimately related to the fact that we may subject the charges to a Lorentz transformation

$$\tilde{I}_i = U(\Lambda)^+ I_i(t)\, U(\Lambda) = \int_\Sigma d\sigma_\mu j_i^\mu(x)\,; \qquad \Sigma : (\Lambda x)^0 = t \tag{2.6}$$

where Σ denotes the Lorentz transform of the surface $x^0 = t$. Clearly the charges $\tilde{I}_i$ also satisfy the algebra (2.2) and therefore satisfy a sum rule of the type (2.5). On the other hand the matrix elements of $\tilde{I}_i$ and $I_i(t)$ are related by [3]

$$\langle p', \alpha' | \tilde{I}_i | p, \alpha \rangle = \langle \Lambda p', \alpha' | I_i(t) | \Lambda p, \alpha \rangle\,. \tag{2.7}$$

The sum rule obtained by sandwiching the Lorentz transformed generators between states of momentum p and p' is therefore identical with the sum rule obtained by considering the original generators between states of momenta Λp and $\Lambda p'$. Different choices of the initial momentum $\boldsymbol{p}$ are therefore equivalent to different choices of the spacelike surface Σ upon which the charges are defined.

3. Infinite Momentum

Fubini and *Furlan* [4] have proposed a particular choice of the initial momentum $\boldsymbol{p}$ in sum rules of the type (2.5), the so-called infinite momentum limit. According to this prescription the momentum $\boldsymbol{p}$ is taken to tend to infinity along a particular direction, usually chosen as the direction of the third axis. As a result only one particular member of the family of sum rules contained in (2.5) is picked out. We shall for the moment adopt this prescription as a recipe and shall return to its virtues later on.

It is instructive to translate the infinite momentum limit to coordinate space by recalling that different choices of the momenta in the sum rule (2.5) are equivalent to different choices of the surface Σ in (2.6). In order to find the surface corresponding to infinite momentum we refer to Eq. (2.7) and observe that the limit may be achieved by making use of a pure Lorentz transformation Λ in the direction of the third axis. If the velocity v characterizing this transformation tends to the velocity of light, the third component of the momentum Λp tends to infinity as prescribed. The corresponding charge $\tilde{I}_i$ is an integral over a surface Σ which approaches the light cone as $v \to c$. If the original spacelike surface is characterized by $x^0 = 0$, the resulting limiting surface is given by the equation $x^0 + x^3 = 0$.

$$\hat{I}_i = \lim_{v \to c} U^+(\Lambda)\, I_i(0)\, U(\Lambda) = \int_{x^0 + x^3 = 0} d\sigma_\mu\, j_i^\mu(x)\,. \tag{3.1}$$

Since the surface $x^0 + x^3 = 0$ has a lightlike normal $n_\mu = (1, 0, 0, 1)$ we shall refer to $\hat{I}_i$ as a lightlike charge. More generally we define

$$\hat{I}_i(\tau) = \int_{n_\mu x^\mu = \tau} d\sigma_\mu j_i^\mu(x)\,; \qquad n_\mu = (1, 0, 0, 1)\,. \tag{3.2}$$

The essential point which has been noted independently by various authors [5] is the following. The assumption that the algebra (2.2) is satisfied only in the infinite momentum limit is equivalent to the hypothesis that the lightlike charges $\hat{I}_i(\tau)$ satisfy the algebra

$$[\hat{I}_i(\tau), \hat{I}_k(\tau)] = i C_{ikl} \hat{I}_l(\tau) \tag{3.3}$$

with no restriction on the choice of initial and final momenta. In fact, if the lightlike algebra (3.3) is sandwiched between one-particle states of arbitrary momentum there does not result a family of sum rules. Instead there results a single one, namely the one corresponding to the Fubini-Furlan-limit of (2.5). The fact that the algebra of the lightlike charges is weaker than the algebra of the conventional spacelike charges may appear to be paradoxical at first sight. The following simple observation provides an intuitive explanation. Lorentz invariance implies the validity of the algebra (2.2) on every spacelike surface if it is valid on a single one. Likewise the algebra (3.3) must hold on every lightlike surface if it is valid on a single one. We are thus actually comparing two *sets* of algebras. The spacelike set is more powerful simply because there are more spacelike surfaces than lightlike ones.

We shall refer to the octet of lightlike charges which satisfy the Lie algebra of $SU(3)$ as $SU(3)_l$. Analogously $SU(2)_l \times SU(2)_l$ denotes the algebra generated by isospin and by the triplet of axial charges contained in a lightlike surface.

It may be worth mentioning that the celebrated Adler-Weisberger-relation is obtained by making use of the infinite momentum limit, i.e. its validity checks $SU(2)_l \times SU(2)_l$ rather than conventional $SU(2) \times SU(2)$.

These considerations may be summarized as follows. Current algebra is an effort to establish a link between observables – the currents – and quantities with simple algebraic properties – the generators of a Lie algebra. The moral of the success of the infinite momentum limit is that this link should be modified by identifying the elements of the algebra not with the charges contained in a spacelike surface but rather with the charges contained in a lightlike surface. Note that the two types of charges coincide if the current is conserved.

4. Virtues of the Lightlike Charges

Many of the practical difficulties that arise if one tries to saturate current algebra sum rules without making use of the infinite momentum limit are an immediate consequence of *Coleman*'s theorem [6] which runs as follows. Let $I_i(t)$ denote a spacelike charge defined by (1.1) in terms of a local current. Coleman has shown that if this charge leaves the vacuum invariant, $I_i(t)|0\rangle = 0$, then the operator $I_i(t)$ commutes with the Hamiltonian and therefore generates a symmetry – "the invariance of the vacuum is the invariance of the world."

This theorem does not apply to lightlike charges. On the contrary the lightlike charges $\hat{I}_i(\tau)$ automatically leave the vacuum invariant whether or not they generate a symmetry [7]. This may be verified as follows. The component $P_{\|} = n_\mu P^\mu = P^0 + P^3$ of the total momentum generates translations within the lightlike surface $n_\mu x^\mu = \tau$. Therefore this quantity commutes with $\hat{I}_i(\tau)$ [8]. This implies that the vector $\hat{I}_i(\tau)|0\rangle$ is an eigenstate of $P_{\|}$ with eigenvalue zero. On the other hand, the spectrum of the operator $P_{\|} = P^0 + P^3$ is evidently positive and the eigenvalue zero can arise only if the state in question has zero mass. Since the vacuum is the only strongly interacting zero mass state we conclude that the operators $\hat{I}_i(\tau)$ map the vacuum onto itself. Moreover, since the vacuum expectation value of the currents vanishes (Lorentz invariance) we have the result $\hat{I}_i(\tau)|0\rangle = 0$. For the lightlike algebras of the type $SU(3)_l$ or $SU(2)_l \times SU(2)_l$ etc. the invariance of the vacuum is not the invariance of the world. This simple fact explains why the infinite momentum limit is such a convenient device. The difficulties that arise with spacelike charges at finite momentum as corollaries of Coleman's theorem are avoided.

A very interesting question arises in this context. It is well-known that the conventional generators $I_i(t)$ defined as integrals over a spacelike surface do not exist as operators, if they do not annihilate the vacuum, i.e. if the corresponding currents are not conserved [9]. This means that the basic commutation rules of conventional charge algebra cannot be viewed as operator relations unless the algebra represents a symmetry of the Hamiltonian [10]. The lightlike charges on the other hand do annihilate the vacuum. Under what conditions do they exist as operators? More work is needed to give an answer to this question.

A further virtue of the lightlike charges has been known since the invention of the infinite momentum limit is the fact that they carry zero invariant momentum transfer. This is because the lightlike charges commute with the parallel component $P_{\|} = P^0 + P^3$ as well as with the transverse components P^1 and P^2 of the total momentum. Therefore the difference of initial and final momenta $q = p' - p$ in a matrix element of the type $\langle p', \alpha' | \hat{I}_i(\tau) | p, \alpha \rangle$ satisfies $q^1 = q^2 = q^0 + q^3 = 0$ which implies $t = (q^0)^2 - (q^1)^2 - (q^2)^2 - (q^3)^2 = 0$. These matrix elements therefore describe the form factors associated with the underlying currents at zero momentum transfer.

5. Quark Model

It is instructive to illustrate the properties of the lightlike charges by means of the free quark model defined by the field equation

$$- i\gamma^\mu \partial_\mu \psi^a + m_a \psi^a = 0 \qquad (a = 1, 2, 3). \tag{5.1}$$

The masses m_1, m_2, and m_3 will be taken to satisfy $m_1 = m_2 \neq m_3$ such that the model is invariant with respect to $SU(2)$ but not with respect to $SU(3)$. The conventional expansion of the field in terms of creation and annihilation operators reads

$$\begin{aligned}\psi^a(x) = (2\pi)^{-3/2} \sum_s \int d^3 p \sqrt{\frac{m_a}{p^0}} \{ & a(\boldsymbol{p}, s, a)\, u(\boldsymbol{p}, s, a)\, e^{-ipx} \\ & + b^+(\boldsymbol{p}, s, a)\, v(\boldsymbol{p}, s, a)\, e^{ipx} \}.\end{aligned} \tag{5.2}$$

The canonical spinors associated with momentum $\boldsymbol{p}$ are defined by means of a pure Lorentz transformation in the direction of $\boldsymbol{p}$

$$u(\boldsymbol{p}, s, a) = S[\Lambda(p)]\, u(s) \tag{5.3}$$

where the quantity $S[\Lambda]$ stands for the 4×4 representation of the homogeneous Lorentz group and $u(s)$ denotes the spinor in the rest frame.

In the following we shall however make use of a different definition of the spinors which is more appropriate for dealing with lightlike surfaces. First of all it is convenient to label the momenta by the parallel component

$p_{\|} = p^0 + p^3$ and by the transverse components $\underline{p} = (p^1, p^2)$ instead of the canonical set $\boldsymbol{p} = (p^1, p^2, p^3)$. The full expression for the momentum in terms of these quantities and the mass reads

$$p^\mu = \{(2p_{\|})^{-1}(p_{\|}^2 + \underline{p}^2 + m^2), \underline{p}, (2p_{\|})^{-1}(p_{\|}^2 - \underline{p}^2 - m^2)\}\,. \tag{5.4}$$

In order to generate an arbitrary momentum characterized by $(p_{\|}, \underline{p})$ from the rest frame $(p_{\|} = m, \underline{p} = 0)$ we first apply a boost in the direction of the third axis with hyperbolic angle α.

$$B(\mathrm{e}^\alpha) = \exp(-i\alpha N_3)\,. \tag{5.5}$$

This transformation takes the rest momentum into $(p_{\|} = \mathrm{e}^\alpha m, \underline{p} = 0)$ and we may choose the angle α such that $p_{\|}$ takes the prescribed (positive) value. We then apply a Lorentz transformation which leaves the value of $p_{\|}$ unchanged, but affects the transverse components. In order not to change the value of $p_{\|} = n_\mu p^\mu$ this Lorentz transformation must belong to the little group of the lightlike vector n_μ. This little group is generated by J_3 and by the operators F_1, F_2 defined by [11]

$$F_1 = N_1 + J_2\,; \qquad F_2 = N_2 - J_1 \tag{5.6}$$

in terms of the conventional generators $\boldsymbol{J} = (M_{23}, M_{31}, M_{12})$ and $\boldsymbol{N} = (M_{10}, M_{20}, M_{30})$ of the homogeneous Lorentz group. The operator J_3 leaves the momentum $(p_{\|}, \underline{p} = 0)$ invariant. To generate transverse momentum use of the operators F_1 and F_2 has to be made. It is straightforward to verify that the Lorentz transformation

$$F(\underline{\beta}) = \exp - i(\beta_1 F_1 + \beta_2 F_2) \tag{5.7}$$

takes the momentum $(p_{\|}, \underline{p} = 0)$ into $(p_{\|}, \underline{p} = \underline{\beta} p_{\|})$. The combined transformation $F(\underline{p}/p_{\|})\, B(p_{\|}/m)$ therefore transforms the rest frame $(p_{\|} = m, \underline{p} = 0)$ into a momentum with parallel and transverse components $p_{\|}$ and $\underline{p}$ respectively. The full expression for this momentum is given in (5.4).

We now replace the pure Lorentz transformation $\Lambda(p)$ in the definition (5.3) of the canonical spinors by the transformation $F(\underline{p}/p_{\|})\, B(p_{\|}/m_a)$ and define a new spinor $\hat{u}(p_{\|}, \underline{p}, s, a)$ by

$$\hat{u}(p_{\|}, \underline{p}, s, a) = S[F(\underline{p}/p_{\|})\, B(p_{\|}/m_a)]\, u(s)\,. \tag{5.8}$$

This spinor is again a solution of the Dirac equation with momentum p. It differs from the canonical spinor by a rotation of the spin direction, since the two Lorentz transformations $F(\underline{p}/p_{\|})\, B(p_{\|}/m_a)$ and $\Lambda(p)$ differ by an element of the rotation group

$$\begin{aligned} F(\underline{p}/p_{\|})\, B(p_{\|}/m_a) &= \Lambda(p)\, R(p) \\ \hat{u}(p_{\|}, \underline{p}, s, a) &= \sum_{s'} u(\boldsymbol{p}, s', a)\, D[R(p)]_s^{s'}\,. \end{aligned} \tag{5.9}$$

The quantity $D[R]_s^{s'}$ denotes the spin 1/2 representation of the rotation group.

In terms of the spinors $\hat{u}(p_{\|}, \underline{p}, s, a)$ the plane wave expansion of the free field may be written as

$$\psi^a(x) = (2\pi)^{-3/2} \sum_s \int \mathrm{d}p_{\|}\, \mathrm{d}^2\underline{p} \sqrt{\frac{m_a}{p_{\|}}}\, \{\hat{a}(p_{\|}, \underline{p}, s, a)\, \hat{u}(p_{\|}, \underline{p}, s, a)\, \mathrm{e}^{-ipx} + \hat{b}^+(p_{\|}, \underline{p}, s, a)\, \hat{v}(p_{\|}, \underline{p}, s, a)\, \mathrm{e}^{ipx}\}\,. \tag{5.10}$$

The new annihilation operators $\hat{a}(p_{\|}, \underline{p}, s, a)$ are related to the canonical quantities by

$$\hat{a}(p_{\|}, \underline{p}, s, a) = \sum_{s'} a(\boldsymbol{p}, s', a)\, D[R^{-1}(p)]_s^{s'} (p^0/p_{\|})^{1/2} \tag{5.11}$$

and satisfy the commutation rules

$$[\hat{a}(p_{\|}, \underline{p}, s, a), \hat{a}^+(p'_{\|}, \underline{p}', s', a')]_+ = \delta(p'_{\|} - p_{\|})\, \delta(\underline{p}' - \underline{p})\, \delta_{s's}\, \delta_{a'a}\,. \tag{5.12}$$

The negative frequency spinors $\hat{v}(p_{\|}, \underline{p}, s, a)$ and the corresponding creation operators $\hat{b}^+(p_{\|}, \underline{p}, s, a)$ are introduced in an entirely analogous fashion.

This completes our discussion of the plane wave expansion of the field $\psi^a(x)$ and we now turn to the evaluation of the lightlike charges associated with the model. The currents are defined as usual by

$$\begin{aligned} j_i^\mu(x) &= \tfrac{1}{2} :\overline{\psi}(x)\, \lambda_i \gamma^\mu \psi(x):\,, \\ a_i^\mu(x) &= \tfrac{1}{2} :\overline{\psi}(x)\, \lambda_i \gamma^\mu \gamma_5 \psi(x):\,. \end{aligned} \tag{5.13}$$

Inserting the plane wave expansion (5.10) in these currents and carrying out the integrals

$$\begin{aligned} \hat{I}_i(0) &= \int_{x^0 + x^3 = 0} \mathrm{d}\sigma_\mu j_i^\mu(x) \\ \hat{A}_i(0) &= \int_{x^0 + x^3 = 0} \mathrm{d}\sigma_\mu a_i^\mu(x) \end{aligned} \tag{5.14}$$

one obtains

$$\begin{aligned} \hat{I}_i(0) &= \tfrac{1}{2} \int \mathrm{d}p_{\|}\, \mathrm{d}^2\underline{p}\, \lambda_{ib}^a \sum_s \{\hat{a}^+(p_{\|}, \underline{p}, s, a)\, \hat{a}(p_{\|}, \underline{p}, s, b) \\ &\quad - \hat{b}^+(p_{\|}, \underline{p}, s, b)\, \hat{b}(p_{\|}, \underline{p}, s, a)\}\,, \\ \hat{A}_i(0) &= \tfrac{1}{2} \int \mathrm{d}p_{\|}\, \mathrm{d}^2\underline{p}\, \lambda_{ib}^a \sum_s \varepsilon_s \{\hat{a}^+(p_{\|}, \underline{p}, s, a)\, \hat{a}(p_{\|}, \underline{p}, s, b) \\ &\quad + \hat{b}^+(p_{\|}, \underline{p}, s, b)\, \hat{b}(p_{\|}, \underline{p}, s, a)\} \end{aligned} \tag{5.15}$$

where $\varepsilon_\uparrow = 1$, $\varepsilon_\downarrow = -1$.

The most remarkable property of these expressions is that they do not contain any cross terms of the type $\hat{a}^+\hat{b}^+$ or $\hat{a}\hat{b}$. Of course this property reflects the fact that the lightlike charges annihilate the vacuum. [Note that the spacelike positive parity charges $I_i(t)$ do contain cross terms unless the masses of the three quarks coincide, i.e. unless the operators $I_i(t)$ are constants of the motion. Even in this symmetry limit the spacelike negative parity charges $A_i(t)$ still contain cross terms. They disappear only if all three quarks have zero mass.]

Finally we observe that the annihilation operators transform according to

$$\begin{aligned} [\hat{I}_i(0), \hat{a}(p_{\|}, \underline{p}, s, a)] &= -\tfrac{1}{2}\lambda^a_{ib}\hat{a}(p_{\|}, \underline{p}, s, b)\,, \\ [\hat{A}_i(0), \hat{a}(p_{\|}, \underline{p}, s, a)] &= -\tfrac{1}{2}\lambda^a_{ib}\varepsilon_s\hat{a}(p_{\|}, \underline{p}, s, b)\,. \end{aligned} \tag{5.16}$$

This shows that *despite the mass splitting the three quarks constitute a triplet representation of the group* $SU(3)_l \times SU(3)_l$.

This is of course not true for the spacelike charges for which the right hand side of (5.16) contains a linear combination of creation and annihilation operators of the type $\alpha\hat{a} + \beta\hat{b}^+$.

6. Local Transformation Properties

We now consider the transformation properties of the quark field itself. With respect to the conventional spacelike charges the field transforms according to

$$\left.\begin{aligned} [I_i(t), \psi^a(x)] &= -\tfrac{1}{2}\lambda^a_{ib}\psi^b(x) \\ [A_i(t), \psi^a(x)] &= -\tfrac{1}{2}\lambda^a_{ib}\gamma_5\psi^b(x) \end{aligned}\right. \quad (x^0 = t)\,. \tag{6.1}$$

One might expect that the same equations hold if $I_i(t)$ and $A_i(t)$ are replaced by $\hat{I}_i(\tau)$ and $\hat{A}_i(\tau)$ respectively provided that x lies on the surface $n_\mu x^\mu = \tau$. This is however not correct. Instead we only have the weaker statements

$$\left.\begin{aligned} [\hat{I}_i(\tau), \psi^a_{\|}(x)] &= -\tfrac{1}{2}\lambda^a_{ib}\psi^b_{\|}(x) \\ [\hat{A}_i(\tau), \psi^a_{\|}(x)] &= +\tfrac{1}{2}\lambda^a_{ib}\gamma_5\psi^b_{\|}(x) \end{aligned}\right. \quad (n_\mu x^\mu = \tau)\,, \tag{6.2}$$

where $\psi^a_{\|}(x)$ denotes the projection

$$\begin{aligned} &\psi^a_{\|}(x) = \gamma_{\|}\psi^a(x)\,, \\ &\gamma_{\|} = n_\mu\gamma^\mu = \gamma^0 + \gamma^3\,; \quad \gamma^2_{\|} = 0\,. \end{aligned} \tag{6.3}$$

This means that only two out of the four components of the spinor $\psi^a(x)$ satisfy a simple local transformation law. This peculiarity is related to the fact that the initial value problem for a lightlike surface is quite different from the canonical initial value problem for a spacelike surface. In the latter case one may prescribe all four components of the spinor $\psi^a(x)$ independently at time $x^0 = t$ say. The Dirac equation then determines the solution for $x^0 > t$. It is however not possible to prescribe all four components independently on a lightlike surface. To see why this is not allowed we rewrite the Dirac equation in terms of the coordinates $\tau = x^0 + x^3, \sigma = x^0 - x^3$ and $\underline{x} = (x^1, x^2)$

$$-i\{\gamma_{\|}\,\partial_\tau + \gamma\,\partial_\sigma + \underline{\gamma}\cdot\underline{\partial}\}\,\psi^a + m_a\psi^a = 0\,. \tag{6.4}$$

The quantity γ stands for $\gamma^0 - \gamma^3$. The matrix $\gamma_{\|}$ multiplying the derivative with respect to τ is singular and the term $\partial_\tau\psi^a$ disappears if we multiply the equation with the projection operator $\frac{1}{2}\gamma\gamma_{\|} = 1 + \gamma^0\gamma^3$. Therefore the spinor $\psi^a(x)$ must satisfy the *constraint*

$$4i\,\partial_\sigma\gamma\psi^a = \gamma\{+i\underline{\gamma}\cdot\underline{\partial} + m_a\}\,\psi^a_{\|} \tag{6.5}$$

which involves only derivatives within the surface $n_\mu x^\mu = \tau$. This shows that the quantity $\gamma\psi^a$ cannot be specified independently of $\psi^a_{\|}$ on the surface $n_\mu x^\mu = \tau$.

The fact that the projection $\psi^a_{\|}(x)$ suffices to specify the solution completely is also evident from the inversion formula

$$\hat{a}(p_{\|}, \underline{p}, s, a) = (2\pi)^{-3/2}\sqrt{\frac{m_a}{p_{\|}}}\int_{n_\mu x^\mu = \tau} d\sigma_\mu\, e^{ipx}\hat{\bar{u}}(p_{\|}, \underline{p}, s, a)\,\gamma^\mu\psi^a(x) \tag{6.6}$$

whose right hand side involves only $\psi^a_{\|}(x)$ since the surface element $d\sigma_\mu$ is proportional to the lightlike vector n_μ.

The situation is quite analogous for the transformation law of the currents. For the spacelike charges we have the local transformation rule

$$[I_i(t), j^\mu_k(x)] = i f_{ikl}\, j^\mu_l(x) \qquad (x^0 = t) \tag{6.7}$$

and analogous relations hold for axial currents and axial charges. In the case of the lightlike charges however, only the components $j_{i\|}(x) = n_\mu j^\mu_i(x)$ and $a_{i\|}(x) = n_\mu a^\mu_i(x)$ transform in this fashion

$$\left.\begin{aligned}
[\hat{I}_i(\tau), j_{k\|}(x)] &= i f_{ikl}\, j_{l\|}(x)\\
[\hat{I}_i(\tau), a_{k\|}(x)] &= i f_{ikl}\, a_{l\|}(x)\\
[\hat{A}_i(\tau), j_{k\|}(x)] &= i f_{ikl}\, a_{l\|}(x)\\
[\hat{A}_i(\tau), a_{k\|}(x)] &= i f_{ikl}\, j_{l\|}(x)
\end{aligned}\right\}\quad (n_\mu x^\mu = \tau)\,. \tag{6.8}$$

In the quark model neither the transverse components (j_k^1, j_k^2) and (a_k^1, a_k^2) nor the longitudinal components $j_k^0 - j_k^3$ and $a_k^0 - a_k^3$ transform according to the regular representation under $SU(3)_l \times SU(3)_l$.

7. Weinberg Sum Rules

As an application of the transformation rules discussed in the last section we now derive *Weinberg*'s sum rules [12] in the present framework. Let us denote the octet of vector currents that characterize the hadronic part of weak and electromagnetic interactions by $j_i^\mu(x)$ and let us denote by $\hat{I}_i(\tau)$ the corresponding set of lightlike charges. We assume that these charges satisfy the commutation rules of $SU(3)_l$ and that the components $j_{i\parallel}(x) = j_i^0(x) + j_i^3(x)$ of the currents transform like octets under this group. This assumption was shown to be satisfied in the framework of the free quark model in the last section. We claim that if it is satisfied in the real world, then Weinberg's first sum rule must hold.

To derive the sum rule we consider the vacuum expectation value of two currents, $\langle 0|j_{i\parallel}(x)\, j_{k\parallel}(y)|0\rangle$, and observe that if x and y are contained in the same lightlike surface, the product $j_{i\parallel}(x)\, j_{k\parallel}(y)$ transforms according to the representation $\underline{8} \times \underline{8}$ under the group $SU(3)_l$ generated by the lightlike charges that belong to this surface. On the other hand, since the lightlike charges leave the vacuum invariant, only the invariant contained in $\underline{8} \times \underline{8}$ contributes

$$\langle 0|j_{i\parallel}(x)\, j_{k\parallel}(y)|0\rangle = \delta_{ik} f(x-y) \qquad (n_\mu x^\mu = n_\mu y^\mu)\,. \tag{7.1}$$

The spectral functions are defined by

$$\begin{aligned}&\langle 0|j_i^\mu(x)\, j_k^\nu(y)|0\rangle \\ &\quad = (2\pi)^{-3} \int d^4p\, \theta(p^0)\, e^{-ipz} \{(p^\mu p^\nu/p^2 - g^{\mu\nu})\, \underset{1}{\varrho_{ik}}(p^2) + p^\mu p^\nu \underset{0}{\varrho_{ik}}(p^2)\}\end{aligned} \tag{7.2}$$

where $z = x - y$. We specialize this general expression to the case that the points x and y are contained in the same lightlike surface, i.e. $n_\mu z^\mu = z^0 + z^3 = 0$. In this case the phase pz is independent of the mass; it involves only the components $p_\parallel$ and $\underline{p}$ of the momentum, $pz = p_\parallel z^0 - \underline{p} \cdot \underline{z}$. Transforming the variables p^0, p^1, p^2, p^3 to the equivalent set $m^2, p_\parallel, \underline{p}$

$$d^4p = (2p_\parallel)^{-1}\, dm^2\, dp_\parallel\, d^2\underline{p} \tag{7.3}$$

the integrand therefore factorizes as follows

$$\begin{aligned}\langle 0|j_{i\parallel}(x)\, j_{k\parallel}(y)|0\rangle &= R_{ik} \tfrac{1}{2} (2\pi)^{-3} \int dp_\parallel\, d^2\underline{p}\, p_\parallel\, e^{-ipz} \\ R_{ik} &= \int dm^2 \{m^{-2} \underset{1}{\varrho_{ik}}(m^2) + \underset{0}{\varrho_{ik}}(m^2)\}\end{aligned} \tag{7.4}$$

comparing this result with (7.1) we obtain the first Weinberg spectral function sum rule

$$R_{ik} = \delta_{ik} S \tag{7.5}$$

where S is some unknown constant.

Two points are to be noted in the present derivation of the sum rule: If one would replace the lightlike charges by spacelike generators and *assume* that the vacuum is invariant under this set one would arrive at a condition similar to (7.1) valid however for $x^0 = y^0$ instead of $n_\mu x^\mu = n_\mu y^\mu$. Comparison with the spectral representation would in this case immediately imply that both spectral functions are proportional to δ_{ik}, i.e. *full symmetry of the spectral functions.* It is remarkable that the group $SU(3)_l$ implies only the much weaker *sum rule* (7.5) for the spectral functions.

The present derivation avoids the use of local current-current commutation rules which were needed in the original derivation by *Weinberg.*

To close this section we observe that the second Weinberg sum rule may be derived in an entirely analogous fashion starting from the requirement that the transverse components $j_i^1(x)$ and $j_i^2(x)$ of the current also transform like an octet. Note however that this requirement is not satisfied in the quark model.

The extension of the calculation to the axial currents, i.e. to the algebras $SU(2)_l \times SU(2)_l$ or $SU(3)_l \times SU(3)_l$ is straightforward.

References

1. *Gerstein, S. S.*, and *J. B. Zeldovitch:* Zh. Eksper. Teor. Fiz. **29**, 698 (1955); — *Feynman, R. P.*, and *M. Gell-Mann:* Phys. Rev. **109**, 193 (1958).
2. For an excellent account of current algebra theory see *Renner, B.:* Current algebras and their applications. New York: Pergamon Press 1968.
3. To simplify the argument we assume that the initial and final states have spin zero.
4. *Fubini, S.*, and *G. Furlan:* Physics **1**, 229 (1965).
5. *Bardakci, K.*, and *G. Segré:* Phys. Rev. **159**, 1263 (1967); — *Susskind, L.:* Phys. Rev. **165**, 1535 (1967); — *Stern, J.:* Dubna Preprint E2-3469 (1967); — *Jersak, J.*, and *J. Stern:* Dubna Preprint E2-3990 (1968); — *Bardakci, K.*, and *M. B. Halpern:* Theories at Infinite Momentum, Lawrence Radiation Lab., University of California Preprint UCRL-18360 (1968); — *Leutwyler, H.:* Lecture presented at the VII. Internationale Universitätswochen, Schladming (1968).
6. *Coleman, S.:* J. Math. Phys. **7**, 787 (1966).
7. *Jersak, J.*, and *J. Stern:* Dubna Preprint E2-3990 (1968).
8. The same is, of course, true of the transverse components P^1 and P^2, whereas, unless we ignore mass splitting, neither the Hamiltonian P^0 nor the operator $M^2 = P_\mu P^\mu$ commute with $\hat{I}_i(\tau)$. Instead we have the equation of motion

$$\frac{d}{d\tau}\hat{I}_i(\tau) = i(2P_{||})^{-1}[M^2, \hat{I}_i(\tau)] .$$

9. *Coleman, S.:* Phys. Letters **19**, 144 (1965); — *Fabri, E.*, and *L. E. Picasso:* Phys. Rev. Letters **16**, 408 (1966); — *Schroer, B.*, and *P. Stichel:* Commun. Math. Phys. **3**, 258 (1966).
10. For an interpretation of these commutation rules in terms of linear forms see Ref. [9].
11. A more detailed discussion of the properties of this little group is contained in the papers quoted in Ref. [5].
12. *Weinberg, S.:* Phys. Rev. Letters **18**, 507 (1967); — *Glashow, S. L., Howard J. Schnitzer*, and *S. Weinberg:* Phys. Rev. Letters **19**, 139 (1967); — *Das, T., V. S. Mathur*, and *S. Okubo:* Phys. Rev. Letters **18**, 761 (1967).

Prof. Dr. *H. Leutwyler*
Institut für Theoretische Physik der Universität
CH - 3000 Bern, Sidlerstraße 5

Introduction to the Lagrangian Method

VOLKHARD F. MÜLLER

Contents

1. Introduction

The following two lectures are presented for purely pedagogical reasons. They contain nothing new and their only purpose is to provide an elementary introduction into the Lagrangian formalism of classical field theory. We shall not deal with the complications arising in the quantized version of the theory, where the field functions become operator-valued distributions. The idea underlying the Lagrangian method is to obtain the equations of motion (i.e. in our case the field equations) as Euler equations of a certain variational principle. Clearly the specific field equations which one desires dictate the choice of the Lagrangian. One merit of the Lagrangian method is to provide the conservation laws corresponding to symmetries of the interaction in a systematic way, known as *Noether's* theorem [1].

We are interested in the local interaction of various types of wave fields ϕ_A transforming covariantly with respect to the inhomogeneous Lorentz group. The index A accounts both for the various types of fields and for the different components of a multicomponent field (e.g. a spinor field or a vector field). Fields corresponding to charged particles are complex. The Lagrangian density is assumed to be a real polynomial of the field components and their first derivatives:

$$L(x) := L\left(\phi_A, \phi_A^*, \frac{\partial \phi_A}{\partial x_\lambda}, \frac{\partial \phi_A^*}{\partial x_\lambda}\right), \tag{1}$$

such, that it transforms as a scalar under inhomogeneous Lorentz transformations.

Comments:

1. A real Lagrangian density is taken, in order to give a self adjoint Hamiltonian in the quantized form.

2. $L(\phi'(x'), \ldots) = L(\phi(x), \ldots)$, gives relativistically covariant field equations.

3. First derivatives of the field functions give differential equations of second order at most.

4. Instead of $\mathrm{Re}\,\phi_A$ and $\mathrm{Im}\,\phi_A$, ϕ_A and ϕ_A^* are considered as independent variables.

The field equations are the Euler equations of the variational principle

$$\int_{\Sigma_1}^{\Sigma_2} \mathrm{d}^4 x\, L(x) = \text{extremum}\,, \tag{2}$$

where Σ_1 and Σ_2 are two space-like surfaces and the variations of the field variables are arbitrary but have to vanish on the boundary. This gives

$$\frac{\partial L}{\partial \phi_A} = \frac{\partial}{\partial x_\lambda} \frac{\partial L}{\partial \dfrac{\partial \phi_A}{\partial x_\lambda}}\,; \quad \frac{\partial L}{\partial \phi_A^*} = \frac{\partial}{\partial x_\lambda} \frac{\partial L}{\partial \dfrac{\partial \phi_A^*}{\partial x_\lambda}}\,. \tag{3}$$

The correspondence between Lagrangian density and field equations is not unique. If we add to the Lagrangian density the divergence of a four-vector function of the field variables

$$L'(x) = L(x) + \frac{\partial}{\partial x_\lambda} F_\lambda(\phi_A, \phi_A^*) \tag{4}$$

the field equations are not altered.

2. Noether's Theorem

We are now going to discuss Noether's theorem for the case of an arbitrary internal symmetry group, i.e. transformations not acting on the spacetime coordinates of the fields. As examples of internal symmetry groups we just mention baryon- or hypercharge gauge transformations, isospin-symmetry or $SU(3)$-symmetry. Given any compact Lie-group G with the corresponding Lie-algebra

$$\mathfrak{G} = \{F^k\} \qquad k = 1, 2, \ldots, L \tag{5}$$

we have

$$[F^r, F^s] = i c_{rst} F^t \quad \text{(summation convention)}. \tag{6}$$

In order not to be bothered by too many different types of indices, we consider the case of two different multiplets of fields, each transforming according to an irreducible (unitary) representation of the compact Lie-group G.

$$\psi = \begin{pmatrix} \psi^1 \\ \vdots \\ \psi^m \end{pmatrix}, \quad \phi = \begin{pmatrix} \phi^1 \\ \vdots \\ \phi^n \end{pmatrix}, \tag{7}$$

$$\psi'(x) := e^{-i \sum_{k=1}^{L} \alpha^k M^k} \psi(x); \quad \phi'(x) := e^{-i \sum_{k=1}^{L} \alpha^k N^k} \phi(x), \tag{8}$$

$$\begin{aligned} &M^r : m \times m \text{ matrices}; \quad M^r = (M^r)^+; \quad [M^r, M^s] = i c_{rst} M^t, \\ &N^r : n \times n \text{ matrices}; \quad N^r = (N^r)^+; \quad [N^r, N^s] = i c_{rst} N^t. \end{aligned} \tag{9}$$

Furthermore the multiplet ϕ shall be real. With respect to inhomogeneous Lorentz transformations all components within a multiplet transform in the same way.

The Lagrangian density then depends on

$$L\left(\psi, \psi^*, \frac{\partial \psi}{\partial x_\lambda}, \frac{\partial \psi^*}{\partial x_\lambda}, \phi, \frac{\partial \phi}{\partial x_\lambda}\right). \tag{10}$$

From $L(\psi', \psi'^*, \ldots)$ we calculate

$$\begin{aligned} \left(\frac{\partial L}{\partial \alpha_k}\right)_{\alpha=0} = \frac{1}{i} \Bigg\{ &\frac{\partial L}{\partial \psi} M^k \psi + \frac{\partial L}{\partial \frac{\partial \psi}{\partial x_\lambda}} M^k \frac{\partial \psi}{\partial x_\lambda} \\ &- \psi^* M^k \frac{\partial L}{\partial \psi^*} - \frac{\partial \psi^*}{\partial x_\lambda} M^k \frac{\partial L}{\partial \frac{\partial \psi^*}{\partial x_\lambda}} \\ &+ \frac{\partial L}{\partial \phi} N^k \phi + \frac{\partial L}{\partial \frac{\partial \phi}{\partial x_\lambda}} N^k \frac{\partial \phi}{\partial x_\lambda} \Bigg\}. \end{aligned}$$

Using the field equations, we can write.

$$\left(\frac{\partial L}{\partial \alpha_k}\right)_{\alpha=0} = \frac{\partial}{\partial x_\lambda} j_\lambda^k(x) \tag{11}$$

with

$$j_\lambda^k(x) := \frac{1}{i}\left\{\frac{\partial L}{\partial\frac{\partial\psi}{\partial x_\lambda}} M^k\psi - \psi^* M^k \frac{\partial L}{\partial\frac{\partial\psi^*}{\partial x_\lambda}} + \frac{\partial L}{\partial\frac{\partial\phi}{\partial x_\lambda}} N^k\phi\right\}. \tag{12}$$

Now, if the Lagrangian density is invariant with respect to the internal symmetry group, that is

$$L(\psi', \psi^{*\prime}, \ldots) = L(\psi, \psi^*, \ldots) \tag{13}$$

we have

$$\left(\frac{\partial L}{\partial\alpha_k}\right)_{\alpha=0} = 0 \tag{14}$$

and thus

$$\frac{\partial j_\lambda^k}{\partial x_\lambda} = 0 \qquad k = 1, 2, 3, \ldots, L. \tag{15}$$

Any compact Lie-group parametrized by L real parameters, acting as an internal symmetry group and *leaving invariant* the Lagrangian density, provides L conserved real currents! Therefore the corresponding L generalized charges

$$F^k := \int d^3x j_0^k(x) \tag{16}$$

are time-independent, that is $\dot{F}^k = 0$.

Instead of coordinate-independent parameters we could have considered group parameters $\alpha^k = \alpha^k(x)$ depending on the space-time coordinates.

$$\psi'(x) = e^{-i\alpha^k(x)M^k}\psi(x), \qquad \phi'(x) = e^{-i\alpha^k(x)N^k}\phi(x). \tag{17}$$

We have seen the invariance of the Lagrangian under internal symmetry transformations with coordinate-independent group parameters α^k provides L conserved currents. If we now consider coordinate-dependent group parameters $\alpha^k(x)$ the Lagrangian will no longer be invariant under transformations with such parameters. Nevertheless, because of

$$\frac{\partial\psi'}{\partial x_\lambda} = -i\frac{\partial\alpha^k(x)}{\partial x_\lambda} M^k\psi'(x) + e^{i\alpha^k(x)M^k}\frac{\partial\psi}{\partial x_\lambda}$$

and similar equations for $\frac{\partial\psi^{*\prime}}{\partial x_\lambda}$ and $\frac{\partial\phi'}{\partial x_\lambda}$ we could have calculated the conserved currents from coordinate-dependent group transformations by

$$j_\lambda^k(x) = \left(\frac{\partial L(\Psi', \Psi^{*\prime}, \Phi')}{\partial\frac{\partial\alpha^k(x)}{\partial x_\lambda}}\right)_{\alpha^k=0;\,\frac{\partial\alpha^k}{\partial x_\lambda}=0}. \tag{18}$$

3. Electrodynamics and Gauge Transformations

Before studying internal symmetry groups with space-time dependent parameters, let us have a short look on spinor electrodynamics as a simple example of the ideas presented.

$$L(x) = \frac{i}{2}\left\{\overline{\psi}(x)\,\gamma_\lambda \frac{\partial \psi(x)}{\partial x_\lambda} - \frac{\partial \overline{\psi}(x)}{\partial x_\lambda}\gamma_\lambda \psi(x)\right\} - M_0 \overline{\psi}(x)\,\psi(x) + e\overline{\psi}(x)\,\gamma^\mu \psi(x)\,A_\mu(x) - \frac{1}{4}F_{\mu\nu}(x)\,F^{\mu\nu}(x) \tag{19}$$

with

$$F_{\mu\nu} := \frac{\partial A_\mu}{\partial x^\nu} - \frac{\partial A_\nu}{\partial x^\mu}. \tag{20}$$

Consider the gauge transformations of the first kind

$$\psi' = \mathrm{e}^{-i\alpha}\psi, \qquad \overline{\psi}' = \overline{\psi}\,\mathrm{e}^{i\alpha}, \qquad A'_\mu = A_\mu. \tag{21}$$

The Lagrangian density (19) is invariant under these transformations

$$L(\psi', \overline{\psi}', \ldots) = L(\psi, \overline{\psi}, \ldots). \tag{22}$$

As a consequence we have the conserved current

$$j_\lambda = \overline{\psi}\gamma_\lambda\psi, \qquad \frac{\partial j_\lambda}{\partial x_\lambda} = 0 \tag{23}$$

either calculated using Eq. (12) or Eq. (18).
We could add a mass term to the Lagrangian (19)

$$L' = L + \frac{\mu_0^2}{2}\cdot A_\lambda(x)\,A^\lambda(x) \tag{24}$$

and still would have the conserved current, as this additional term is invariant under gauge transformations of the first kind.

But if we use gauge transformations of the second kind:

$$\psi'(x) = \mathrm{e}^{-ie\Lambda(x)}\psi(x)\,; \qquad \overline{\psi}'(x) = \overline{\psi}(x)\,\mathrm{e}^{ie\Lambda(x)}$$
$$A'_\mu(x) = A_\mu(x) - \frac{\partial \Lambda(x)}{\partial x^\mu} \tag{25}$$

the Lagrangian density (19) is invariant under these transformations (25), but not the Lagrangian density (24). The mass term breaks the symmetry. The free Lagrangian of the vector potential, $-\frac{1}{4}F^{\mu\nu}F_{\mu\nu}$, is invariant under gauge transformations of the second kind. In order that the full Lagrangian density is invariant under such transformations, the differential operators acting on the Dirac fields and the vector potential A_λ have to enter only in the combination

$$\frac{\partial}{\partial x^\lambda} - ie\,A_\lambda \qquad \text{«minimal coupling»}\,. \tag{26}$$

We can restate this situation in the following way: The Lagrangian density for the Dirac field is invariant under gauge transformations of the first kind but not under gauge transformations of the second kind. If one desires invariance under gauge transformations of the second kind one is forced to introduce the vector potential according to (26) and transforming as given by Eq. (25).

The vector potentials A_λ have to obey the Lorentz condition $\frac{\partial A_\lambda}{\partial x_\lambda} = 0$. This restricts the gauge functions $\Lambda(x)$ to solutions of the equation $\Box \Lambda(x) = 0$. Both types of gauge transformations form abelian groups.

4. Generalized Yang-Mills Theories [2]

The generators of an internal symmetry group commute with the generators of the inhomogeneous Lorentz-group. *Yang* and *Mills* took up this point and required that the Lagrangian density of strong interactions should be invariant under independent isospin transformations at all space-time points. They constructed the general Lagrangian density having this symmetry. Their theory is a generalization of the gauge invariance of the second kind encountered in electrodynamics to a non-abelian gauge group.

We are going to discuss the method of Yang and Mills for any compact simple Lie-group as internal symmetry. The corresponding Lie-algebra $\mathfrak{G} = \{F^k\}, k = 1, 2 \ldots, L$ has a hermitian basis, $F^k = (F^k)^+$, and fulfils

$$[F^r, F^s] = i c_{rst} F^t \tag{27}$$

with structure constants c_{rst} having the properties

$$c_{rst} \qquad \text{real and totally antisymmetric}\,, \tag{28 a}$$

$$c_{rmk}\,c_{stm} + c_{smk}\,c_{trm} + c_{tmk}\,c_{rsm} = 0\,. \tag{28 b}$$

We consider two types of fields, ψ^k and ϕ^l, each transforming according to an irreducible unitary representation of the internal symmetry group. [See formulae (7), (8) and (9), but with $\alpha^k = \alpha^k(x)$.] In the following we shall call an internal symmetry transformation with space-time dependent parameters $\alpha^k(x)$ an internal symmetry gauge transformation (ISGT).

The infinitesimal ISGT are then

$$\begin{aligned} \psi'(x) &= (1 - i g_0 \alpha^k(x) M^k)\, \psi(x) \\ \phi'(x) &= (1 - i g_0 \alpha^k(x) N^k)\, \phi(x) . \end{aligned} \tag{29}$$

From this follows

$$\begin{aligned} \frac{\partial \psi'(x)}{\partial x_\lambda} &= (1 - i g_0 \alpha^k(x) M^k) \frac{\partial \psi}{\partial x_\lambda} - i g_0 \frac{\partial \alpha^k(x)}{\partial x_\lambda} M^k \psi(x) , \\ \frac{\partial \phi'(x)}{\partial x_\lambda} &= (1 - i g_0 \alpha^k(x) N^k) \frac{\partial \phi}{\partial x_\lambda} - i g_0 \frac{\partial \alpha^k(x)}{\partial x_\lambda} N^k \phi(x) . \end{aligned} \tag{30}$$

In order to construct a Lagrangian density invariant under ISGT, we start with a Lagrangian density

$$L_0 \left(\psi, \psi^*, \frac{\partial \psi}{\partial x_\lambda}, \frac{\partial \psi^*}{\partial x_\lambda}, \phi, \frac{\partial \phi}{\partial x_\lambda} \right) \tag{31}$$

invariant under transformations of the internal symmetry group with space-time-independent parameters. This leads, as we have seen [Eq. (11) and (12)] to L conserved currents

$$j_\lambda^k = \frac{1}{i} \left\{ \frac{\partial L_0}{\partial \dfrac{\partial \psi}{\partial x_\lambda}} M^k \psi - \psi^* M^k \frac{\partial L_0}{\partial \dfrac{\partial \psi^*}{\partial x_\lambda}} + \frac{\partial L_0}{\partial \dfrac{\partial \phi}{\partial x_\lambda}} N^k \phi \right\} \quad k = 1, 2, \ldots, L . \tag{32}$$

If we now use coordinate dependent parameters, i.e. perform an ISGT, L_0 will no longer be invariant under those transformations, as can be seen from Eq. (30) for infinitesimal parameters. In order to compensate the additional terms containing the derivatives of the parameters we introduce L real vector fields $v_\lambda^k(x)$ transforming according to the adjoint representation of the internal symmetry group. Under an infinitesimal ISGT they are to transform as

$$v_\lambda'^k(x) = v_\lambda^k(x) - \frac{\partial \alpha^k(x)}{\partial x^\lambda} + g_0 c_{klm} \alpha^l(x)\, v_\lambda^m(x) . \tag{33}$$

The extra term $g_0 c_{klm} \alpha^l v^m$ which is absent in gauge transformations of the second kind in electrodynamics, originates from the non-abelian nature of the symmetry group.

If we define

$$\begin{aligned} D_\lambda &:= \frac{\partial}{\partial x^\lambda} - i g_0 \sum_{k=1}^{L} M^k v_\lambda^k , \\ d_\lambda &:= \frac{\partial}{\partial x^\lambda} - i g_0 \sum_{k=1}^{L} N^k v_\lambda^k \end{aligned} \tag{34}$$

then the Lagrangian density

$$L_0(\psi, \psi^*, D_\lambda \psi, (D_\lambda \psi)^*, \phi, d_\lambda \phi) \tag{35}$$

is invariant under ISGT. This follows immediately from the fact that under an infinitesimal ISGT we have

$$\begin{aligned} D'_\lambda \psi'(x) :&= \left(\frac{\partial}{\partial x_\lambda} - i g_0 \sum_{k=1}^{L} M^k v'^k_\lambda \right) \psi'(x) \\ &= (1 - i g_0 \alpha^k(x) M^k) D_\lambda \psi(x) \end{aligned}$$

and a similar relation for $d'_\lambda \phi'(x)$. In order to get the full Lagrangian for the 3 fields ψ, ϕ and v_λ invariant under ISGT, we have to add a term L_v containing the kinetic energy part of the v_λ fields which is itself invariant under ISGT. Defining

$$G^k_{\mu\nu}(x) : = \frac{\partial v^k_\mu(x)}{\partial x^\nu} - \frac{\partial v^k_\nu(x)}{\partial x^\mu} - g_0 c_{klm} v^l_\mu(x)\, v^m_\nu(x) \tag{36}$$

we see from Eq. (33) that this quantity transforms under an infinitesimal ISGT as

$$G'^k_{\mu\nu}(x) = G^k_{\mu\nu}(x) + g_0 c_{klm} \alpha^l(x)\, G^m_{\mu\nu}(x)\,. \tag{37}$$

Thus we can choose

$$L_v : = -\tfrac{1}{4} G^k_{\mu\nu} G^{k\mu\nu} \tag{38}$$

which is invariant under ISGT.

Now the Lagrangian density

$$L : = L_0(\psi, \psi^*, D_\lambda \psi, (D_\lambda \psi)^*, \phi, d_\lambda \phi) + L_v \tag{39}$$

describes the system of fields ψ, ϕ and v_λ interacting with each other in such a way L is invariant under internal symmetry gauge transformations.

Since the Lagrangian density (39) is invariant under ISGT, it is a forteriori invariant under transformations with constant parameters α^k. This invariance leads, as we have seen, to L conserved currents

$$J^k_\lambda(x) = j^k_\lambda(x) - c_{klm} G^l_{\lambda\mu}(x)\, v^{m\mu}(x) + \text{possibly additional terms} \tag{40}$$

coming from $L_0(\psi, D_\lambda \psi, \ldots) - L_0(\psi, \partial_\lambda \psi, \ldots)$ $k = 1, 2, \ldots, L$,

$$\frac{\partial J^k_\lambda(x)}{\partial x_\lambda} = 0\,. \tag{41}$$

We see, that the original currents j^k_λ [see Eq. (32)] are no longer conserved.

If we would add a mass term

$$+\frac{\mu_0^2}{2} v_\lambda^l(x)\, v^l(x)^\lambda \tag{42}$$

to the Lagrangian density (39), this term would violate the invariance under ISGT, but not under transformations with constant parameters and thus providing the conserved currents (40) too. A remarkable implication of the invariance under ISGT is the universal coupling of the vector fields v_λ^k to the fields ψ and ϕ. In case of invariance under symmetry transformations with constant parameters only, this universality would not be implied.

As in the case of electrodynamics we have a restriction on the gauge functions $\alpha^k(x)$ if we require that they preserve the spin-one condition

$$\frac{\partial v_\lambda^k}{\partial x_\lambda} = 0\,. \tag{43}$$

From Eq. (33) we immediately derive this restriction to be

$$\Box \alpha^k(x) = g_0 c_{klm} \frac{\partial \alpha^l(x)}{\partial x_\lambda} v_\lambda^m(x)\,. \tag{44}$$

The result obtained can be generalized to a compact *semi-simple* internal symmetry group. Since a semi-simple Lie-algebra is the direct sum of a finite number of simple ideals [3], completely independent Yang-Mills fields can be introduced with respect to each simple ideal.

We illustrate the ideas presented with the isospin-invariant meson-nucleon interaction.

$$SU(2): [I^k, I^l] = i\varepsilon_{klm} I^m\,, \tag{45}$$

$$\psi'(x) = \left(1 - \frac{i}{2}\vec{\alpha}\vec{\tau}\right)\psi(x)\,, \qquad (t^k)_{lm} = \frac{1}{i}\varepsilon_{klm}\,.$$

$$\phi'(x) = (1 - i\vec{\alpha}\vec{t})\,\phi(x)\,,$$

The isospin-invariant Lagrangian is

$$\begin{aligned} L_0\left(\psi, \bar{\psi}, \frac{\partial \psi}{\partial x}, \frac{\partial \bar{\psi}}{\partial x}, \phi, \frac{\partial \phi}{\partial x}\right) :&= \frac{i}{2}\left\{\bar{\psi}\gamma \frac{\partial \psi}{\partial x} - \frac{\partial \bar{\psi}}{\partial x}\gamma\psi\right\} - M_0 \bar{\psi}\psi \\ &+ \frac{1}{2}\frac{\partial \vec{\phi}}{\partial x_\lambda}\cdot\frac{\partial \vec{\phi}}{\partial x^\lambda} - \frac{1}{2} m_0^2\, \vec{\phi}\cdot\vec{\phi} + G\bar{\psi}\gamma_5 \vec{\tau}\psi\cdot\vec{\phi}\,. \end{aligned} \tag{46}$$

This Lagrangian density leads to the conserved currents

$$\vec{j}_\lambda := \bar{\psi}\gamma_\lambda \frac{\vec{\tau}}{2}\psi + \left(\vec{\phi} \times \frac{\partial \vec{\phi}}{\partial x^\lambda}\right) \tag{47}$$

and to the field equations

$$\left(-i\gamma\frac{\partial}{\partial x}+M_0\right)\psi=G\vec{\tau}\gamma_5\psi\vec{\phi}\,,\qquad(48)$$
$$(\Box+m_0^2)\,\vec{\phi}=G\bar{\psi}\gamma_5\vec{\tau}\psi\,.$$

Construction of the isospin-gauge-invariant Lagrangian density L:

$$D_\lambda:=\frac{\partial}{\partial x^\lambda}-ig_0\frac{\vec{\tau}}{2}\cdot\vec{v}_\lambda\,;\qquad d_\lambda:=\frac{\partial}{\partial x^\lambda}-ig_0\vec{t}\cdot\vec{v}_\lambda\,,\qquad(49)$$

$$L_0(\psi,\bar{\psi},D_\lambda\psi,\overline{D_\lambda\psi},\phi,d_\lambda\phi)=L_0\left(\psi,\bar{\psi},\frac{\partial\psi}{\partial x_\lambda},\frac{\partial\bar{\psi}}{\partial x_\lambda},\phi,\frac{\partial\phi}{\partial x'_\lambda}\right)$$
$$+g_0\vec{v}_\lambda\cdot\vec{j}^\lambda+\frac{g_0^2}{2}(\vec{v}_\lambda\times\vec{\phi})\cdot(\vec{v}^\lambda\times\vec{\phi})\,,\qquad(50)$$

$$\vec{G}_{\mu\nu}:=\frac{\partial\vec{v}_\mu}{\partial x^\nu}-\frac{\partial\vec{v}_\nu}{\partial x^\mu}-g_0(\vec{v}_\mu\times\vec{v}_\nu)\,,\qquad(51)$$

$$L_v=-\frac{1}{2}\frac{\partial\vec{v}_\mu}{\partial x^\nu}\cdot\frac{\partial\vec{v}^\mu}{\partial x_\nu}+\frac{1}{2}\frac{\partial\vec{v}_\nu}{\partial x^\mu}\cdot\frac{\partial\vec{v}^\mu}{\partial x_\nu}+g_0\frac{\partial\vec{v}^\mu}{\partial x_\nu}\cdot(\vec{v}_\mu\times\vec{v}_\nu)$$
$$-\frac{g_0^2}{4}(\vec{v}_\mu\times\vec{v}_\nu)\cdot(\vec{v}^\mu\times\vec{v}^\nu)\,,\qquad(52)$$

$$L=L_0(\psi,\bar{\psi},D_\lambda\psi,\overline{D_\lambda\psi},\phi,d_\lambda\phi)+L_v\,.$$

From this Lagrangian density follow the conserved isospin-currents

$$\vec{J}_\lambda:=\vec{j}_\lambda+g_0\vec{\phi}\times(\vec{v}_\lambda\times\vec{\phi})-(\vec{v}^\nu\times\vec{G}_{\nu\mu})\,,\qquad(53)$$

$$\frac{\partial\vec{J}_\lambda}{\partial x_\lambda}=0\,,\qquad(54)$$

and the field-equations

$$\left(-i\gamma\frac{\partial}{\partial x}+M_0\right)\psi=G\vec{\tau}\gamma_5\psi\vec{\phi}+g_0\frac{\vec{\tau}}{2}\gamma^\lambda\psi\vec{v}_\lambda\,,$$
$$(\Box+m_0^2)\,\vec{\phi}=G\bar{\psi}\gamma_5\vec{\tau}\psi+2g_0\left(\frac{\partial\vec{\phi}}{\partial x_\lambda}\times\vec{v}_\lambda\right)$$
$$+g_0^2\vec{v}_\lambda\times(\vec{\phi}\times\vec{v}^\lambda)+g_0\left(\vec{\phi}\times\frac{\partial\vec{v}_\lambda}{\partial x_\lambda}\right),\qquad(55)$$
$$\frac{\partial}{\partial x_\nu}\vec{G}_{\nu\mu}=g_0\vec{J}_\mu\,.$$

The last term in the expression (51) for $\vec{G}_{\mu\nu}$ gives rise to nonlinear field-equations for the vector-fields.

References

1. For a general discussion see *Hill, E. L.:* Rev. Mod. Phys. **23**, 253 (1951).
2. *Yang, C. N.*, and *R. L. Mills:* Phys. Rev. **96**, 191 (1954). — *Glashow, S. L., and M. Gell-Mann:* Ann. Phys. (N.Y.) **15**, 437 (1961). — *Utiyama, R.:* Phys. Rev. **101**, 1597 (1956). — *Zumino, B.:* Proc. of the IV. Internationale Universitätswochen für Kernphysik, Schladming 1965, p. 212.
3. See for example: *Chevally, C.:* Théorie des Groupes de Lie, tome III, p. 64—68. Paris: Hermann & Cie 1955.

Dr. *Volkhard F. Müller*
Institut für Theoretische Physik
der Universität
6900 Heidelberg, Philosophenweg 16

Introduction to the Method of Current Algebra

H. PIETSCHMANN

Contents

1. Introduction

Current Algebra is "en vogue" for more than three years now; despite many pessimistic forecasts, it still proves capable of delivering a lot of fruitful results and it still forms one of the fundaments, strong interaction physics is built on.

After a general outline, we shall demonstrate the power of current algebra on two different but outstanding examples, the Adler-Weisberger relation and the Mathur-Okubo-Pandit-Callan-Treiman relation. In the last lecture we shall touch upon sum rules for spectral functions.

2. General Survey of the Method

In current algebra commutation relations in the widest sense are used together with information on the divergence of some kind of current.

The way of application of these two ingredients varies greatly between practical cases. The type of commutation relation used will be mostly those for the generators of some sort of symmetry group. In order to demonstrate the method, we shall calculate in detail an almost trivial example which, however, shows all the necessary ingredients of a full-fledged problem in the field of current algebra. Later on we shall only point out differences to this demonstration calculation when we come to more complex cases. The reader who also asks for the details of the more complex cases is referred to Ref. [1].

The best established internal symmetry group is the well known group of isospin rotations, characterized by the commutation relation

$$[I_+, I_-] = I_3 \tag{1}$$

and two more, in which we are not interested at present. The generators for the isospin group are connected to the time components of the isovector current operators by

$$I_i = \int d^3x j_0^i(x), \quad i = +, -, 3. \tag{2}$$

The isovector currents, in turn, are connected to the experimentally accessible isovector form factors by taking the matrix element between one nucleon states [2]

$$\langle p'|j_\mu^i(x)|p\rangle = \frac{1}{(2\pi)^3} \frac{M}{\sqrt{p_0 p_0'}} \bar{u}(p')\{\gamma_\mu F_1^v(q^2) + i\sigma_{\mu\nu} q^\nu F_2^v(q^2)\} \frac{\tau_i}{2} u(p) e^{iqx}, \tag{3}$$

where

$$q_\mu = p_\mu' - p_\mu \tag{4}$$

is the four momentum transfer. Integration over all space of the time-component of Eq. (3) yields (q_0 vanishes with $|\boldsymbol{q}| \to 0$ as long as the nucleon masses are degenerate)

$$\langle p', \alpha| I_i |p, \beta\rangle = \tfrac{1}{2} F_1^v(0)\, (\tau_i)_{\alpha\beta}\, \delta^{(3)}(\boldsymbol{p}' - \boldsymbol{p}), \tag{5}$$

where we have exhibited the isospin character of the nucleon states ($\alpha = 1$ ... proton, $\alpha = 2$ ... neutron). It is clear that the matrix element of Eq. (1) between one nucleon states will directly give measurable quantities on the right hand side. Using proton states we have

$$\langle p'|[I_+, I_-]|p\rangle = \tfrac{1}{2} F_1^v(0)\, \delta^{(3)}(\boldsymbol{p}' - \boldsymbol{p}). \tag{6}$$

In order to calculate the left hand side of Eq. (6) a complete set of intermediate states is now introduced.

$$\sum_n \{\langle p'|I_+|n\rangle \langle n|I_-|p\rangle - \langle p'|I_-|n\rangle \langle n|I_+|p\rangle\} = \tfrac{1}{2} F_1^v(0)\, \delta^{(3)}(\boldsymbol{p}' - \boldsymbol{p}). \tag{7}$$

At this point, the information on the divergence of the current has to be used to handle the infinite sum over intermediate states. Neglecting electromagnetic effects, the three isovector currents are conserved. Equivalently, this means that all isospin generators are time independent, hence that they are diagonal together with the Hamiltonian except for possible degeneracies in total energy. Therefore, only one nucleon states do contribute to the infinite sum of Eq. (7).

In general, if the underlying group is an exact symmetry, a multiplet corresponding to an irreducible representation of the group will contribute to the infinite sum and the "sum rule" is said to be "saturated" by that multiplet. Otherwise, there is "leakage" from the multiplet.

In the case under consideration only the one neutron state is accessible and thus the second term in the curly bracket drops. The rest becomes

$$\sum_{\text{spins}} \int d^3q \langle p'|I_+|q\rangle \langle q|I_-|p\rangle = \tfrac{1}{2} F_1^v(0)\, \delta^{(3)}(\boldsymbol{p}'-\boldsymbol{p}), \tag{8}$$

where $|q\rangle$ is a simple neutron state. Using Eq. (5) and working out the trivial spin algebra yields

$$\tfrac{1}{2}[F_1^v(0)]^2\, \delta^{(3)}(\boldsymbol{p}'-\boldsymbol{p}) = \tfrac{1}{2} F_1^v(0)\, \delta^{(3)}(\boldsymbol{p}'-\boldsymbol{p})$$

and excluding the trivial solution it follows that

$$F_1^v(0) = 1\,. \tag{9}$$

Hence the physically important result is shown that the "charge is not renormalized by strong interactions". This result is usually derived by means of the *Ward* identity [3] and, historically, it was the first application of current algebra [4] after its suggestion by *Gell-Mann* in 1962 [5].

For a proper understanding, a physical concept has to be clarified. It is well known [3] that the charge is, indeed, renormalized by electromagnetic interactions. We have excluded them by assuming that only the nucleon states are degenerate in energy. In reality, a one nucleon state plus an arbitrary number of soft photons is, of course, also degenerate with the one nucleon state. Hence photons have to be excluded to arrive at Eq. (9). In fact, it is the long range nature of the electromagnetic field (i.e. the existence of soft photons) that causes the renormalization of charge by electromagnetic interactions.

Strictly speaking, we have shown Eq. (9) only for the isovector charge. By a somewhat altered calculation the same equation is shown to hold for the total charge too [1].

3. The Adler-Weisberger Sum Rule

The result (9) is interesting but not new. A similar quantity which has resisted a theoretical calculation for more than a decade is the analogous form factor of the axial vector current, defined by

$$\begin{aligned}\langle p'|j_\mu^{5,i}(x)|p\rangle = \frac{1}{(2\pi)^3}\frac{M}{\sqrt{p_0 p_0'}}\,\bar{u}(p')\,\{\gamma_\mu G_1(q^2) \\ + q_\mu G_2(q^2)\}\,\gamma_5 \frac{\tau_i}{\sqrt{2}}\, u(p)\cdot e^{iqx}.\end{aligned} \tag{10}$$

In complete analogy to Eq. (2) one can define an "isotopic chirality" by

$$Q_a^i = \int d^3x j_0^{5,i}(x) \tag{11}$$

and its matrix element between one nucleon states will be

$$\langle p'\alpha | Q_a^i | p\beta \rangle = \frac{1}{\sqrt{2}} G_1(0)(\tau_i)_{\alpha\beta}\, \delta^{(3)}(\boldsymbol{p}' - \boldsymbol{p}). \tag{12}$$

It is precisely this matrix element that enters the Gamow-Teller matrix element in the decay of a free neutron. Hence $G_1(0)$ is measurable and is often called "G_A/G_V" or simply λ. Its value is about 1,2 with small fluctuations due to gradual refinements in some experiments.

Theoretically, it is of great importance to understand this relatively simple quantity. There is one far reaching difference to the example of Section 2. The axial vector current cannot be conserved because otherwise the pion would be stable against two-particle decays [6]. Moreover, we have to look for new commutation relations involving the isotopic chiralities. Commutation relations of this kind actually have been suggested by *M. Gell-Mann* [5] in 1962 and later on found more support by the quark model or even any other composite particle model. They are

$$[Q_a^+, Q_a^-] = 2I_3 \tag{13}$$

where the factor 2 stems from the different normalization of the axial vector current (10). There are two new facts as compared to Section 2. Firstly, due to the non-conservation of the axial vector current, there is leakage from the one nucleon states and secondly, the simple neutron state contribution depends on the external momentum for the very same reason. Indeed, simple Dirac algebra yields [1]

$$\sum_{\text{spins}} \int d^3q \langle p' | Q_a^+ | q \rangle \langle q | Q_a^- | p \rangle = \lambda^2\, \delta^{(3)}(\boldsymbol{p}' - \boldsymbol{p}) \left(1 - \frac{M^2}{E^2}\right), \tag{14}$$

where

$$\lambda = G_1(0) \tag{15}$$

and E is the total nucleon energy corresponding to momentum $\boldsymbol{p}$ or $\boldsymbol{p}'$.

$$E = [\boldsymbol{p}^2 + M^2]^{1/2}. \tag{16}$$

Thus, instead of a simple sum rule, one obtains a whole family of sum rules depending on the external parameter E. It has been suggested by *Fubini* and *Furlan* [7] that the sum rule shall be evaluated at $E \to \infty$ because this is the value of minimal leakage. It also turns out to be the best way to handle the problem technically [1].

It remains to evaluate the multi-particle intermediate states. In the example of Section 2, they could be excluded altogether because the current was conserved. In the present case, where the current is not conserved, it is only natural that the divergence of the current will also yield information on the multi-particle intermediate states. Indeed, if

$$\frac{\partial}{\partial x_\mu} j_\mu^{5,i}(x) = a\,\phi^i(x) \tag{17}$$

the Heisenberg equation of motion tells us that

$$\langle\alpha|\,Q_a^i\,|\beta\rangle = \frac{-ia\int d^3x\,\langle\alpha|\,\phi^i(x)\,|\beta\rangle}{E_\alpha - E_\beta}\,, \tag{18}$$

where $\langle\alpha|$ and $|\beta\rangle$ are arbitrary states which are *not* degenerate in energy and use has been made of Eq. (11).

Let us now enter the actual calculation. Just as in Section 2, Eq. (7), we now have

$$\begin{aligned}\lambda^2\,\delta^{(3)}(\boldsymbol{p}'-\boldsymbol{p})\left(1-\frac{M^2}{E^2}\right) &+ \sum\{\langle p'|\,Q_a^+\,|\alpha\rangle\langle\alpha|\,Q_a^-\,|p\rangle \\ &- \langle p'|\,Q_a^-\,|\alpha\rangle\langle\alpha|\,Q_a^+\,|p\rangle\} = \delta^{(3)}(\boldsymbol{p}'-\boldsymbol{p})\,,\end{aligned} \tag{19}$$

where we have singled out the one neutron contribution Eq. (14) so that the sum over α excludes single particle contributions. In order to evaluate the infinite sum over multi-particle intermediate states, we use Eqs. (17) and (18). However, we have not yet specified the right hand side of Eq. (17). According to *Gell-Mann* and *Levy* [8] $\phi^i(x)$ can be taken to be the pion field. This is possible, since the matrix element entering the two body weak pion decay is just the matrix element of the axial vector current between the vacuum and the single pion state.

$$\langle 0|\,j_\nu^{5,i}(x)\,|\pi(q)\rangle = \frac{1}{(2\pi)^{3/2}}\,\frac{i}{\sqrt{2q^0}}\,\mu\,f_\pi q_\nu e^{-iqx}\,, \tag{20}$$

where μ is the pion mass and f_π is related to the pion decay by

$$\Gamma_\pi = G^2\,f_\pi^2\,\frac{\mu^3 m_\mu^2}{8_\pi}\left(1-\frac{m_\mu^2}{\mu^2}\right)^2. \tag{21}$$

Using Eq. (20) in the matrix element of Eq. (17) between the vacuum and a single pion state yields

$$a = \mu^2\,f_\pi\,. \tag{22}$$

The well-known *Goldberger-Treiman* relation [9] allows for a further computation of a,

$$f_\pi = \frac{\sqrt{2M_\lambda}}{gK(0)} \tag{23}$$

where g is the πNN coupling constant and $K(0)$ is the form factor of the πNN vertex at "zero mass" of the pion. That Eq. (17) with a given by Eq. (22) and ϕ being the pion field is valid as an operator equation, i.e. also for pions off their mass shell, is known as "P.C.A.C." or "P.D.D.A.C." It is justified by the good validity of the Goldberger-Treiman relation Eq. (23) with $K(0)$ set equal to unity.

With all this in mind, Eq. (19) becomes

$$\begin{aligned}\frac{1}{\lambda^2} = 1 - \frac{M^2}{E^2} - \sum_\alpha \frac{2M_\mu^4}{g^2K^2(0)} (2\pi)^6\, \delta^{(3)}(\boldsymbol{p} - \boldsymbol{q}\alpha) \frac{1}{(E_\alpha - E)^2} \\ \cdot \{|\langle p|\phi^+(0)|\alpha\rangle|^2 - |\langle p|\phi^-(0)|\alpha\rangle|^2\}\,,\end{aligned} \tag{24}$$

where we have dropped the common δ-function and divided by λ^2. At this point, it is useful to remember the suggestion of Fubini and Furlan to take the limit $E \to \infty$. It is necessary to assume that this limit can be interchanged with the infinite sum over intermediate states. This is an additional, independent assumption.

From the system of variables describing a certain definite intermediate state, the center-of-mass variables can be singled out, leaving "internal" variables only. For convenience, the total energy of the system shall be exhibited by an extra δ-function together with an integration. This gives

$$\begin{aligned}\frac{1}{\lambda^2} = 1 - \lim_{E\to\infty} \frac{2M^2\mu^4}{g^2K^2(0)} \int d^3q_\alpha \int_{M+\mu}^{\infty} dW \sum_{\substack{\alpha \\ \text{int}}} \delta(W - M\alpha)\, \delta^{(3)}(\boldsymbol{q}_\alpha - \boldsymbol{p}) \\ \cdot \frac{(2\pi)^6}{(E_\alpha - E)^2} \{|\langle p|\phi^+(0)|\alpha\rangle|^2 - |\langle p|\phi^-(0)|\alpha\rangle|^2\}\,.\end{aligned} \tag{25}$$

The integral over d^3q_α can be carried out by means of the δ-function and then the limit $E \to \infty$ is easily performed, because

$$E_\alpha = [W^2 + E^2 - M^2]^{1/2}\,,$$

so that

$$\lim_{E\to\infty} E(E_\alpha - E) = \tfrac{1}{2}(W^2 - M^2)\,. \tag{26}$$

Defining Lorentz invariant functions $K^\pm(W)$ by

$$K^\pm(W) = \sum_{\substack{\alpha \\ \text{int}}} \delta(W - M_\alpha)\,(2\pi)^6 \left[\frac{E}{M}\,\frac{E_\alpha}{M_\alpha}\right] |\langle p|\phi^\pm(0)\;\alpha\rangle|^2 \tag{27}$$

the sum rule becomes

$$\frac{1}{\lambda^2} = 1 - \frac{2M^2\mu^4}{g^2K^2(0)} \int_{M+\mu}^{\infty} \mathrm{d}W \frac{4MW}{(W^2-M^2)^2} \{K^+(W) - K^-(W)\}\,. \quad (28)$$

The invariant functions $K^{\pm}(W)$ have been defined in such a way that they are closely related to the total cross-section for scattering of a pion of vanishing mass on a proton at a total energy W

$$\sigma_0^{\pm}(W) = \frac{2\pi M\mu^4}{W^2-M^2} K^{\pm}(W)\,. \quad (29)$$

Insertion into Eq. (28) yields the famous *Adler-Weisberger* relation [10]

$$\frac{1}{\lambda^2} = 1 - \frac{4M^2}{\pi g^2K^2(0)} \int_{M+\mu}^{\infty} \mathrm{d}W \frac{W}{W^2-M^2} \{\sigma_0^+(W) - \sigma_0^-(W)\}\,. \quad (30)$$

It allows for a calculation of $G_1(0)$ (or G_A/G_V) from knowledge about the total cross-sections of off-mass shell pions. The analytic continuation to the real pions introduces some minor uncertainties so that the value for λ becomes

$$|\lambda| = 1.16 - 1.24\,. \quad (31)$$

Notice in passing that the integral in (30) converges because of the Pomeranchuck-theorem which requires the curly bracket to vanish at high energies. The weight factor enhances contributions from low energies, i.e. the resonance region. Indeed, it is the (3/2, 3/2) pion-nucleon resonance at 1236 MeV that causes λ to be bigger than unity because σ_0^+ is stronger enhanced than σ_0^- at the energy of the resonance. Higher resonances damp the value but they cannot outweigh the main contribution. It is seen that one obtains dynamical insight into the way how the value of $G_1(0)$ is composed of its various contributions. This is one of the main goals of elementary particle physics of today.

4. Semi-Leptonic K-Meson Decays

In this section we derive the *Mathur-Okubo-Pandit-Callan-Treiman* relation [11]. It is one of the main applications of the so-called "soft-pion-technique". However, since there is an excellent review on this technique at this summer school [12] we can be very brief with our discussion.

The aim is to relate the hadronic matrix element for K_{l2} decay to that of K_{l3} decay. The calculation proceeds in 3 steps. Firstly, the pion of the final state in K_{l3} decay is taken out of the state by means of the reduction formula. Secondly, the limit of vanishing four momentum is taken for technical reasons. This is why one speaks of "soft pions" in analogy to soft photons. The PCAC relation [Eqs. (17) and (22)] is then used to replace the pion field by the divergence of the axial vector current. After a partial integration with respect to time and a check that surface terms do not contribute, one can use current commutation relations to arrive at

$$\lim_{q'\to 0} (2\pi)^{3/2} \sqrt{2q_0'} \langle \pi^0(q')| S^+ |K^+(q)\rangle = -\frac{1}{f_\pi} \langle 0| S^+ |K^+(q)\rangle , \quad (32)$$

where S^+ is the generator of SU_3 obtained in analogy to Eq. (2) but replacing the isovector current by the $\Delta Q = \Delta S$ strangeness changing weak current. The matrix elements of S^+ can be analyzed in the following way

$$\langle \pi^0(q')| S^+ |K^+(q)\rangle = \frac{1}{\sqrt{4q_0 q_0'}} \{ f_+(q_0' + q_0) + f_-(q_0 - q_0')\} \delta^{(3)}(\mathbf{q} - \mathbf{q}') e^{it(q_0' - q_0)} , \quad (33)$$

$$\langle 0| S^+ |K^+(q)\rangle = (2\pi)^{3/2} \frac{1}{\sqrt{2q_0}} f_K m_K^2 \delta(\mathbf{q}) e^{-itq_0} . \quad (34)$$

Experimentally, f_K is measured to be

$$f_K = 0.070 \pm 0.001 . \quad (35)$$

Insertion into Eq. (32) yields the wanted relation

$$f_+ + f_- = \sqrt{2} \frac{m_K}{\mu} \frac{f_K}{f_\pi} . \quad (36)$$

It should be noted that f_+ and f_- are at the unphysical values of vanishing pion mass. A comparison with experimental results is somewhat difficult at the moment since the ratio f_-/f_+ is not known with confidence. In fact, values from polarization measurements contradict those from the ratio $K_{e3}/K_{\mu 3}$ as well as those from the electron spectrum. If the so far generally accepted positive value of about 1/2 for f_-/f_+ remains to be true, quantitative agreement with Eq. (36) is very good.

5. Sum-Rules for Spectral Functions

There is quite a number of different possibilities to derive sum rules for spectral functions. Here, we shall look into the way suggested by *Das*, *Mathur*, and *Okubo* [13]. It is not only based on very physical assumptions but it also allows for a clear presentation of some of the difficulties.

We begin by defining "propagator functions" for vector and axial vector currents.

$$\Delta^V_{\mu\nu}(q) = \int d^4x\, e^{iqx} \langle 0|\, T\, \{j^i_\mu(x), j^{i+}_\nu(0)\}\, |0\rangle\,, \tag{37a}$$

$$\Delta^A_{\mu\nu}(q) = \int d^4x\, e^{iqx} \langle 0|\, T\, \{j^{5,i}_\mu(x), j^{5,i}_\nu(0)\}\, |0\rangle\,. \tag{37b}$$

Although there are labels for the various charge states of the isovector currents, we do not have to label the propagators for we assume SU_2 to be an exact symmetry. If the symmetry was a wider one, $SU_2 \otimes SU_2$ say, we could also skip the labels V and A. However, this is not so in nature because this symmetry is broken. The underlying assumption of asymptotic symmetry now is that for $q^2 \to \infty$ the symmetry should become exact, i.e. that

$$\lim_{q^2 \to \infty} [\Delta^V_{\mu\nu}(q) - \Delta^A_{\mu\nu}(q)] = 0\,. \tag{38}$$

From now on, everything is just formal manipulation. These manipulations are quite standard for this type of investigations. The first thing is to write down the most general form for the bracket in Eq. (38)

$$\Delta^V_{\mu\nu}(q) - \Delta^A_{\mu\nu}(q) = g_{\mu\nu} F(q^2) + q_\mu q_\nu G(q^2) + g_{\mu 0} g_{\nu 0} H\,. \tag{39}$$

The last term is noncovariant and stems from the fact that we are dealing with the time-ordered product. Eq. (38) requires right away that H vanishes, i.e. that the noncovariant terms are equal for vector and axial vector currents. But there are two more relations:

$$\lim_{q^2 \to \infty} F(q^2) = 0 \tag{40}$$

$$\lim_{q^2 \to \infty} q^2 G(q^2) = 0\,. \tag{41}$$

In order to obtain information out of these equations, we write down the spectral representations for Eqs. (37)

$$\Delta^V_{\mu\nu}(q^2) = -\int_0^\infty ds\, \varrho_V(s) \left(g_{\mu\nu} - \frac{q_\mu q_\nu}{s} \right) \frac{1}{q^2 - s + i\varepsilon} + \text{N.C.T.}\,, \tag{42}$$

where we have not specified the noncovariant terms (N.C.T.). The fact that only one spectral function contributes stems from the conservation of the vector currents. The spectral function is given by

$$\varrho_V(q^2) = \tfrac{1}{3}(2\pi)^3 \sum_n \theta(q)\, \delta^{(3)}(p_n - q) \sum_\mu |\langle 0| j_\mu(0)\, |n\rangle|^2\,. \tag{43}$$

The axial vector is not conserved, but in agreement with P.P.D.A.C. we assume that only the pion contributes to its second spectral function

$$\Delta^A_{\mu\nu}(q^2) = -\int_0^\infty \mathrm{d}s\, \varrho_A(s)\left(g_{\mu\nu} - \frac{q_\mu q_\nu}{s}\right)\frac{1}{q^2 - s + i\varepsilon} - f_\pi^2 \frac{q_\mu q_\nu}{q^2 - \mu^2 + i\varepsilon} + \text{N.C.T.}\,, \tag{44}$$

where f_π is the same as in section 3 and the axial vector spectral function is defined similarly to Eq. (43).

From the spectral representations, we can now read off the functions of Eq. (39) to be

$$F(q^2) = \int_0^\infty \mathrm{d}s\, \frac{\varrho_V(s) - \varrho_A(s)}{s - q^2 - i\varepsilon}\,, \tag{45}$$

$$G(q^2) = \int_0^\infty \mathrm{d}s\, \frac{\varrho_V(s) - \varrho_A(s)}{s(s - q^2 - i\varepsilon)} - \frac{f_\pi^2}{\mu^2 - q^2 - i\varepsilon}\,. \tag{46}$$

It is seen that Eq. (40) is identically fulfilled. On the other hand, Eq. (41) leads to the "1st Weinberg sum rule" [14]

$$\int_0^\infty \mathrm{d}s\, \frac{\varrho_V(s) - \varrho_A(s)}{s} = f_\pi^2\,. \tag{47}$$

In order to derive Eq. (47), it was *not* necessary to assume that the pion mass vanishes.

If one insists in also getting relations out of Eq. (40), one has to assume that the symmetry is reached so fast that also the first moment of Eq. (38) vanishes, i.e. that

$$\lim_{q^2 \to \infty} q^2 [\Delta^V_{\mu\nu}(q) - \Delta^A_{\mu\nu}(q)] = 0\,. \tag{48}$$

In this case one obtains the "2nd Weinberg sum rule" [14]

$$\int_0^\infty \mathrm{d}s [\varrho_V(s) - \varrho_A(s)] = 0\,. \tag{49}$$

However, it was pointed out by *Capra* [15] that it is now a consistency requirement that the pion mass vanishes. This is seen if one expands

Eq. (46)

$$G(q^2) = -\frac{1}{q^2}\left\{\int_0^\infty ds\,\frac{\varrho_V - \varrho_A}{s} - f_\pi^2\right\} - \frac{1}{q^4}\left\{\int_0^\infty ds(\varrho_V - \varrho_A) - \mu^2 f_\pi^2\right\} + O\left(\frac{1}{q^6}\right). \tag{50}$$

One possible, though awkward way out would be to assume different asymptotic behaviour for $F(q^2)$ and $G(q^2)$ that is to say that the symmetry limit is reached in $F(q^2)$ faster than in $G(q^2)$. In this case one obtains directly a logarithmically divergent correction [16] to the pion mass difference as calculated by *Das* et al. [17].

Saturation of the two Weinberg sum rules with zero width resonances yields the result

$$\frac{m_\varrho}{m_A} = \left[1 - \frac{f_\pi^2 m_\varrho^2}{g_\varrho^2}\right]^{1/2}. \tag{51}$$

By means of the relation [18]

$$g_\varrho^2 = 2 f_\pi^2 m_\varrho^2 \tag{52}$$

the right hand side of Eq. (51) is converted into the famous square root of 2.

The same scheme can be easily extended to other groups, SU_3 say. The two Weinberg sum rules then read

$$\int_0^\infty ds\,\frac{\varrho_1^{\alpha\beta}(s)}{s} + \int_0^\infty ds\,\varrho_0^{\alpha\beta}(s) = c\,\delta_{\alpha\beta}, \tag{53}$$

$$\int_0^\infty ds\,\varrho_1^{\alpha\beta}(s) = c'\,\delta_{\alpha\beta}, \tag{54}$$

where α and β are SU_3 indices and the spectral functions are defined by

$$\left(g_{\mu\nu} - \frac{q_\mu q_\nu}{q^2}\right)\varrho_1^{\alpha\beta}(q^2) + q_\mu q_\nu \varrho_0^{\alpha\beta}(q^2) = (2\pi)^3 \sum_n \theta(q)\,\delta^{(3)}(p_n - q)\langle 0|j_\mu^\alpha(0)|n\rangle\langle n|j_\nu^{\beta\dagger}(0)|0\rangle. \tag{55}$$

Saturation of these sum rules with zero width resonances leads to symmetry results. Therefore, it seems to be important to go one step further and take into account the two particle continuum. In the case of the off-diagonal $\omega - \phi$ part of Eq. (53) this leads to satisfactory results [19].

References

1. *Pietschmann, H.:* Selected topics in current algebra. "Lectures in Theor. High En. Phys.", ed. *H. H. Aly:* John Wiley and Sons Ltd, London 1968.
2. *Nilsson, J.*, and *H. Pietschmann:* An introduction to weak interaction physics. McGraw Hill Publ. Corp., N. Y. (in preparation).
3. Cf. any textbook on Quantum Electrodynamics.
4. *Balachandran, A. P.*, and *H. Pietschmann:* Acta Phys. Austriaca **16**, 362 (1963); — Nucl. Phys. **43**, 321 (1963).
5. *Gell-Mann, M.:* Phys. Rev. **125**, 1067 (1962).
6. *Taylor, C.:* Phys. Rev. **110**, 1216 (1958).
7. *Fubini, S.*, and *G. Furlan:* Physics **1**, 229 (1965).
8. *Gell-Mann, M.*, and *M. Levy:* Nuovo Cimento **16**, 705 (1960).
9. *Goldberger, M.*, and *S. Treiman:* Phys. Rev. **110**, 1178 (1958).
10. *Adler, S.:* Phys. Rev. Lett. **14**, 1051 (1965). — *Weisberger, W.:* Phys. Rev. Lett. **14**, 1047 (1965).
11. *Callan, C. G.*, and *S. B. Treiman:* Phys. Rev. Lett. **16**, 153 (1966). — *Mathur, V. S.*, *S. Okubo*, and *L. K. Pandit:* Phys. Rev. Lett. **16**, 371 (1966).
12. *Stech, B.:* Talk at this summer school.
13. *Das, T.*, *V. S. Mathur*, and *S. Okubo:* Phys. Rev. Lett. **18**, 761 (1967).
14. *Weinberg, S.:* Phys. Rev. Lett. **18**, 507 (1967).
15. *Capra, F.:* Orsay-preprint.
16. *Kokott, T.*, *H. Pietschmann*, and *H. Rollnik:* Z. Physik **216**, 65 (1968).
17. *Das, T.*, *G. S. Guralnik*, *V. S. Mathur*, *F. E. Low*, and *J. E. Young:* Phys. Rev. Lett. **18**, 759 (1967).
18. *Kawarabayashi*, and *Suzuki:* Phys. Rev. Lett. **16**, 255 (1966). — *Riazuddin*, and *Fayyazuddin:* Phys. Rev. **147**, 1071 (1966).
19. *Dietz, K.*, and *H. Pietschmann:* Nuovo Cimento **52**, 631 (1967).

Prof. Dr. *Herbert Pietschmann*
Institut für Theoretische Physik
der Universität Wien
A-1090 Wien, Boltzmanngasse 5

S-Matrix Formulation of Current Algebra

H. PILKUHN

Contents

1. Strong Interactions

Current algebra has been developed in the framework of local field theory. However, there also exists an S-matrix formulation. *Mandelstam* [1] has recently shown that the "current algebra results" for strong interactions follow from *Adler*'s self-consistency condition [2] alone, without referring to weak interactions or current commutators. The computation of quantities like G_A/G_V of course requires additional assumptions about weak interactions. Following *Mandelstam*, I shall discuss these in the second half of may talk. The motivation for this approach is the suspicion that the theory works only at low momenta (the current-current picture of weak interaction certainly needs modifications at very high energies). It thus may happen that the local operators do not exist, exept for their matrix elements at low momenta.

Mandelstam also tries to derive Adler's condition from the assumption that the pion is a Toller pole with $|M| = 1$, i.e. a Type III conspirator in the classification of *Freedman* and *Wang* [3]. That derivation is open to doubt, so we shall simply start by taking *Adler*'s condition as a postulate: The strong coupling of a pion to any system of hadrons vanishes in the limit $P_\pi \to 0$, where P_π is the pion 4-momentum.

For equal parities in the initial and final states, the coupling vanishes of course at least as $\sqrt{-t}, t = P_\pi^2$ [e.g. $\bar{u}_{M'}(P')\gamma_5 u_M(P) = -2M\,\delta_{MM'}\sqrt{-t}$]. Therefore, the only new case is that of opposite parities. In particular, we now consider elastic pion-baryon scattering, for which we wish to derive *Weinberg*'s formula [4] for the scattering lengths. We decompose the scattering amplitudes T into a part $T^{(+)}$ for $I = 0$ (and possibly $I = 2$) exchange and a part $T^{(-)}$ for $I = 1$ exchange:

$$T = T^{(+)} + T^{(-)} = T^{(+)} + \frac{1}{2}(P_\pi + P'_\pi)^\mu\, T_\mu^{(-)} \tag{1}$$

where P_π and P'_π are the initial and final pion momenta, and $T^{(+)}$ and $T^{(-)}_\mu$ are still operators in isospace. By the Bose-Einstein symmetry in the t-channel, $T^{(+)}$ is an even function of $(P_\pi + P'_\pi)_\mu$, and $T^{(-)}$ is an odd function of $(P_\pi + P'_\pi)_\mu$. The isospin structure of $T^{(-)}_\mu$ is

$$T^{(-)}_\mu = 2\boldsymbol{I}_\pi \boldsymbol{I}_B T_\mu . \tag{2}$$

$\boldsymbol{I}_\pi$ and $\boldsymbol{I}_B$ are the isospin vector operators for the pion and baryon. For small values of $(P_\pi + P'_\pi)^\mu$, $T^{(+)}$ and $T^{(-)}_\mu$ are approximately independent of $(P_\pi + P'_\pi)^\mu$. In a power series expansion, they correspond to the zeroth and first orders in $(P_\pi + P'_\pi)^\mu$. According to Adler's postulate, the zeroth order vanishes, i.e.

$$T^{(+)} = 0 \quad \text{at} \quad (P_\pi + P'_\pi)^\mu = 0 . \tag{3}$$

Next, we introduce the momentum transfer $Q_\mu = (P_\pi - P'_\pi)_\mu = (P' - P)_\mu$. At zero momentum transfer, T_μ has only a time-component in the lab system, and we can write it in the form $\delta_{\mu 0} \cdot 2ma$, where the baryon mass m comes from the normalization $\bar{u}u = 2m$, and the factor a is a constant. We shall now show that for small P^μ_π, a is a universal constant, independent of the baryon under consideration.

We consider the two cases $P'_\pi = 0$, $Q = (P_\pi - P'_\pi) = P_\pi + P'_\pi$, and $P_\pi = 0$, $Q = (P_\pi - P'_\pi) = -(P_\pi + P'_\pi)$, and apply Adler's condition:

$$T^{(+)} \pm \frac{1}{2} Q^\mu T^{(-)}_\mu = O(Q^2) \tag{4}$$

i.e. each term is proportional to Q^2. Of course, putting one pion momentum equal to zero and keeping the other non-zero means that one of the nucleons is not quite on its mass shell. The rest is now analogous to the derivation of charge conservation from gauge invariance [5]. One considers a larger process, e.g. a collision of two baryons with the additional scattering of a soft pion or the production of a soft pion pair. For small P_π and P'_π, these processes are dominated by the baryon poles:

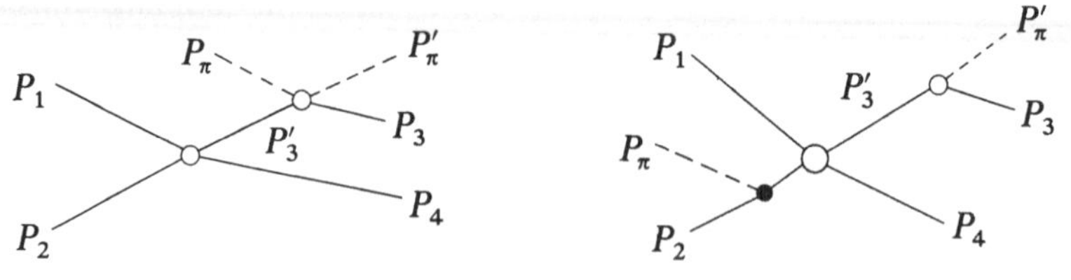

(Actually, the neglect of cuts is an assumption, since for $m_\pi = 0$ the cut branch point coincides with the pole.)

In the first type of poles, both pions are connected to the same external line. For these diagrams, the "divergence condition" $Q^\mu T_\mu^{(-)} = 0$, is

$$Q^\mu T_\mu^{(-)} = 2\boldsymbol{I}_\pi \left[\frac{(P_1' - P_1)(P_1' + P_1)}{m_1^2 - P_1'^2} \boldsymbol{I}_1 a_1 + \frac{(P_2' - P_2)(P_2' + P_2)}{m_2^2 - P_2'^2} \boldsymbol{I}_2 a_2 + \frac{(P_3 - P_3')(P_3 + P_3')}{m_3^2 - P_3'^2} \boldsymbol{I}_3 a_3 + \frac{(P_4 - P_4')(P_4 + P_4')}{m_4^2 - P_4'^2} \boldsymbol{I}_4 a_4 \right] \times T(12 \to 34) = O(Q^2)$$

or

$$\boldsymbol{I}_1 a_1 + \boldsymbol{I}_2 a_2 - \boldsymbol{I}_3 a_3 - \boldsymbol{I}_4 a_4 = 0 . \tag{5}$$

Since this formula must be valid for all possible baryon collisions, one must have $a_1 = a_2 = a_3 = a_4 = a$.

To complete the proof, we must show that the "bremsstrahlung" type of diagrams, in which the two pions are connected to two different external lines, are separately "gauge invariant". Consider for example the emission of a soft pion by particle 3 in the pseudovector coupling:

$$\begin{aligned} T \cong \bar{u}(P_3)\gamma_5 (P_3 - P_3') \cdot \gamma \frac{P_3' \cdot \gamma + m_3}{m_3^2 - P_3'^2} &= \bar{u}(P_3)\gamma_5(-m_3 - \gamma P_3') \frac{P_3' \cdot \gamma + m_3}{m_3^2 - P_3^2} \\ &= \bar{u}(P_3)\gamma_5 \left(1 + 2m_3 \frac{P_\pi \cdot \gamma}{m_3^2 - P_3'^2}\right). \end{aligned} \tag{6}$$

The s-wave contribution tends to zero for $P_\pi \to 0$ (due to the factor γ_5) as expected. The p-wave contribution is a priori undetermined. To be consistent with the Adler condition, we now *define* the limit $P_\pi \to 0$ such that also this contribution (the second term in (6)) vanishes. This is easily done by requiring first $\boldsymbol{p}_\pi \to 0$ (pion rest frame) and then $E_\pi \to 0$. (Recall that the time component of $\bar{u}(P')\gamma_5\gamma_\mu u(P)$ vanishes for $P' \to P$.) The other possibility, a cancellation of various nonvanishing p-wave contributions does not exist, since the p-wave matrix elements are spin dependent, and spin is not conserved.

These results apply to the scattering of massless pions. We shall not discuss possible modifications for physical pions. By crossing symmetry, both $T^{(+)}$ and $T_0^{(-)}$ are even functions of $E_\pi + E_\pi'$. We therefore keep $T^{(+)} = 0$ at the physical threshold. Defining the scattering length a_{sc} by

$$T(E_\pi = m_\pi) = 8\pi m a_{\text{sc}} , \tag{7}$$

we get $a_{\text{sc}} = m_\pi a / 2\pi$. For the scattering length in a state of total isospin I, we get

$$a_{\text{sc}}^I = m_\pi \frac{a}{2\pi} [I(I+1) - I_B(I_B+1) - 1(1+1)] , \tag{8}$$

the square bracket being the eigenvalue of $2\boldsymbol{I}_\pi \boldsymbol{I}_B$. This is Weinberg's formula, with one universal constant a.

2. Weak Interactions

Having derived the necessary properties of the strong interactions, we can now postulate certain properties of the axial current of weak interactions. In S-matrix theory, the matrix elements of the weak currents are defined by the factorization of the leptonic decay amplitudes

$$T = \bar{u}(l')\gamma_\mu(1-\gamma_5)u(l)(V^\mu + A^\mu), \tag{9}$$

where the 4-vectors V^μ and A^μ depend only on the hadron's spins and momenta. The CVC hypothesis reads $\Delta_\mu V^\mu = 0$, with Δ_μ = momentum transfer to the hadrons. From this "gauge invariance" condition, one concludes that V^μ is proportional to the matrix elements of the operators $I_\pm$, again by requiring that the sum of the Born terms of V^μ in more complex weak processes be gauge invariant. The corresponding hypothesis $\Delta_\mu A^\mu = 0$ (CAC) can only be formulated in the limit $P_\pi^2 \to 0$, and if the strong interactions satisfy Adler's condition. For neutron decay, $\Delta_\mu A^\mu = 0$ leads to the Goldberger-Treiman relation (g_π = pion decay constant in the decay matrix element $A^\mu(\pi) = g_\pi \Delta^\mu$)

$$g_\pi = -G_A m / G_{\pi N}, \tag{10}$$

due to the fact that

$$\Delta_\mu \frac{1}{\sqrt{2}} G_A \bar{u}(p)\gamma_5\gamma^\mu u(n) = G_A\sqrt{2}m \neq 0. \tag{11}$$

A pseudoscalar pole $2m\Delta^\mu/t$ ($t = \Delta^2$) must be subtracted from γ^μ to make $\Delta_\mu A^\mu = 0$. This pole obviously corresponds to the pole of the massless pion (Goldstone boson)

$$\sqrt{2}\,G_{\pi N}\bar{u}(p)\gamma_5 u(n)\, 1/t\, g_\pi \Delta^\mu, \tag{12}$$

with $\sqrt{2}\,G_{\pi N}$ = coupling constant of the charged pion to the nucleon.

Next, consider the collision of a neutrino with a neutron, producing an electron, a neutron and a soft pion of momentum P'_π. For $P'_\pi \to 0$, A^μ is of the form

$$A^\mu \cong \frac{G'_A}{\sqrt{2}} \bar{u}(p)\gamma^\mu u(n) - \Delta^\mu \frac{g_\pi}{t} \tilde{T}(\pi^+ n \to \pi^+ n). \tag{13}$$

The presence of the soft pion manifests itself in the first term only by the omission of γ_5 and the replacement of G_A by an unknown factor G'_A. The symbol $\tilde{T}$ indicates that this amplitude is taken at $P_\pi = P'_\pi = 0$. The first term is gauge invariant by itself, $\Delta_\mu \bar{u}(p)\gamma^\mu u(n) = 0$, but the second term gives $-g_\pi \tilde{T}(\pi^+ n \to \pi^+ n)$. Therefore, (13) is automatically gauge-invariant if Adler's condition is fulfilled. Historically, of course, this was the place where the condition was discovered.

Finally, g_π/G_v (or equivalently G_A/G_v) may be determined by requiring that the matrix elements of the commutator between two axial charges enter with the same normalization factor as those of the commutator between two vector charges. The product of two weak currents occurs in the weak scattering of lepton pairs. In particular, we now consider the scattering on nucleons, without the production of additional hadrons.

For $P_\pi \to 0$, the singular parts are given by the two diagrams

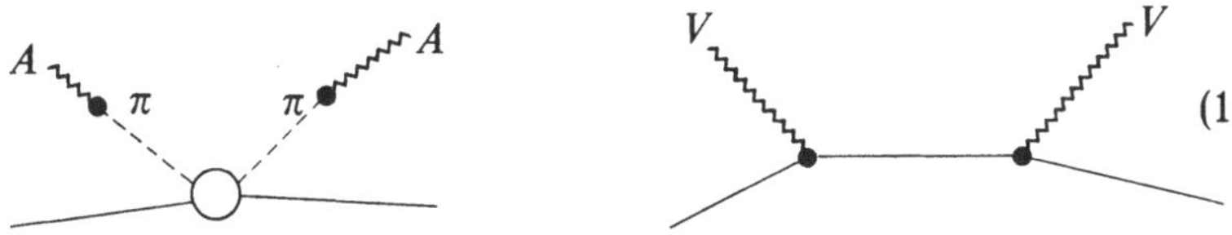

(14)

the matrix elements of which are, for small P,

$$T_{\mu\nu}^{AA} = g_\pi^2 \frac{P_{\pi\mu}}{t} \frac{P'_{\pi\nu}}{t'} \frac{1}{2} (P_\pi + P'_\pi)^\mu T_\mu \cdot 2 I_\pi I_N ,$$
$$T_{\mu\nu}^{VV} = \frac{G_v^2}{2} \bar{u}(P') \gamma_\mu \frac{(P+P_\pi)\cdot\gamma + m}{m^2 - (P+P_\pi)^2} \gamma_\nu u \cdot 2 I_\pi I_N . \tag{15}$$

There is also a nucleon pole in $T_{\mu\nu}^{AA}$, but it does not appear in T_{00}^{AA} (Gamow-Teller transition).

We now put $\boldsymbol{P}_\pi = \boldsymbol{P}'_\pi = 0$ and obtain for small values of $E_\pi = E'_\pi$,

$$T_{\mu\nu}^{AA} = g_\pi^2 \cdot \frac{1}{E_\pi} \cdot 2ma \cdot 2 I_\pi I_N \cdot \delta_{\mu 0} \delta_{\nu 0} ,$$
$$T_{\mu\nu}^{VV} = G_v^2 \cdot \frac{m}{E_\pi} \cdot 2 I_\pi I_N \delta_{\mu 0} \delta_{\nu 0} . \tag{16}$$

Obviously, these two expressions have the same structure. They become identical for

$$g_\pi^2 \cdot 2a = G_v^2 . \tag{17}$$

This is the Adler-Weisberger relation in its low-energy form. Expressing a in terms of the scattering length a_{sc}, and g_π^2 by means of (10), we get the more familiar form of *Weinberg* [4]

$$4\pi \frac{1}{m_\pi} a_{sc} = \frac{G_v^2}{g_\pi^2} = \left(\frac{G_v}{G_A}\right)^2 \frac{g_{\pi N}^2}{m^2} . \tag{18}$$

So far, we have considered just the product of two weak processes. When one considers processes with additional soft external pions, one finds that only the commutator part of the product can be identical for axial vector charges. This completes the "straightbackward" derivation of current algebra.

References

1. *Mandelstam, S.:* Phys. Rev. **168**, 1884 (1968).
2. *Adler, S.:* Phys. Rev. **137** B, 1022 (1964).
3. *Freedman, D. Z.*, and *Jung-Ming Wang:* Phys. Rev. Letters **17**, 569 (1966); — Phys. Rev. **153**, 1596 (1967).
4. *Weinberg, S.:* Phys. Rev. Letters **17**, 616 (1966).
5. — Phys. Rev. **135**, B 1049 (1964).

Dr. *H. Pilkuhn*
Institut für Theoretische Kernphysik der Universität
7500 Karlsruhe 1, Kaiserstraße 12

Electromagnetic Mass Differences

J. ROTHLEITNER

Contents

1. Introduction

In this lecture I will discuss the problem of the shift of the energy levels of strongly interacting systems caused by the electromagnetic interaction. These level shifts are notoriously infinite and this fact is the origin of "mass"-renormalisation. In Section 2 I will shortly discuss the Coulomb-shift of nonrelativistic "nuclear" levels where "mass" renormalisation gives the completely satisfactory solution. It consists in the observation that the number of particles is conserved in such a theory and that we therefore can renormalise the energy of an n-particle system by subtracting n-times the (infinite) self energy of one particle. I shall discuss this in some detail for bound states of 2 particles (deuteron). We will find that the level shift can be expressed in terms of the elastic deuteron form factor and the cross section for inelastic electron deuteron (coulomb) scattering. We shall see that the divergence which has to be removed by "mass"-renormalisation is contained in the inelastic part, whereas the elastic part is always convergent due to the rapid decrease of the form factor which can be proved from the short-distance behaviour of the wave function of the bound state. The slow decrease of the total inelastic electron-deuteron cross-section for large momentum transfer reflects in an experimentally testable relation for the large momentum transfer behaviour of that cross section. It seems that these features largely carry over to the case of the hadrons. In Section 3 I will turn to the electromagnetic self-energies of hadrons. Here sofar we can only calculate the elastic contributions which are convergent due to the rapid decrease of the form factors found experimentally. Whereas these contributions seem to give the bulk of the $I=2$ mass differences the $I=1$ mass differences must be of an essentially different origin. They must be essentially determined by the inelastic contributions which, similar to the nonrelativistic case, lead to infinite mass shifts, but what is much worse, the theory we have sofar leads to infinite mass differences, even in the mild world where *Bjorkens* [1] limit applies. This is discussed in Section 4. Finally in Section 5 a possible connection of the $I=1$ mass differences with nonleptonic weak decays in the framework of the field algebra model [2] is described.

2. The Problem of Nonrelativistic Coulomb-Shift and its Solution by "Mass"-Renormalisation

The problem is characterized by the Hamiltonian $H=H_{\text{st}}+H'$, where the small perturbation H' is given by the Coulomb interaction:

$$H' = \frac{e^2}{2}\int \mathrm{d}^3x\,\mathrm{d}^3y\,\frac{\varrho(\boldsymbol{x})\,\varrho(\boldsymbol{y})}{4\pi|\boldsymbol{x}-\boldsymbol{y}|}\,. \tag{2.1}$$

The divergence of the Coulomb self energy originates in the singular $1/r$ behaviour of the Coulomb potential at $r=0$. For a clean discussion of the "mass"-renormalisation we replace $V(r)=\frac{1}{4\pi r}$ by $V_\Lambda(r):=\frac{1}{4\pi r}-\frac{e^{-\Lambda r}}{4\pi r}$ and take the limit $\Lambda\to\infty$. The renormalised interaction H'_{ren} then becomes

$$\begin{aligned} H'_{\text{ren}} &= \frac{e^2}{2}\int \mathrm{d}^3x\,\mathrm{d}^3y\,\varrho(\boldsymbol{x})\,\varrho(\boldsymbol{y})\,V_\Lambda(|\boldsymbol{x}-\boldsymbol{y}|)-\frac{e^2\Lambda}{8\pi}\int \mathrm{d}^3x\,\varrho(\boldsymbol{x}) \\ &= \int \mathrm{d}^3x\,\mathscr{H}'_{\text{ren}}(\boldsymbol{x}) \end{aligned} \tag{2.2}$$

It is easy to see that now the self energy of a single particle is 0.

Turning now to the shift of the binding energy of the "deuteron"-state $\left|\begin{matrix}\varepsilon_B & j\\ \boldsymbol{p} & s\end{matrix}\right)$ which is an eigenstate of H_{st}:

$$H_{\text{st}}\left|\begin{matrix}\varepsilon_B & j\\ \boldsymbol{p} & s\end{matrix}\right) = \left(\frac{\boldsymbol{p}^2}{4m}-\varepsilon_B\right)\left|\begin{matrix}\varepsilon_B & j\\ \boldsymbol{p} & s\end{matrix}\right) \qquad (m = \text{nucleonmass}) \tag{2.3}$$

we find from 1st order perturbation theory:

$$\delta\varepsilon = \lim_{\Lambda\to\infty}\left(\begin{matrix}\varepsilon_B & j\\ \boldsymbol{p} & s\end{matrix}\middle|\mathscr{H}'_{\text{ren}}(0)\middle|\begin{matrix}\varepsilon_B & j\\ \boldsymbol{p} & s\end{matrix}\right) \tag{2.4}$$

in the normalisation

$$\left(\begin{matrix}\varepsilon_B & j\\ \boldsymbol{p}' & s'\end{matrix}\middle|\begin{matrix}\varepsilon_B & j\\ \boldsymbol{p} & s\end{matrix}\right) = (2\pi)^3\,\delta_{s's}\,\delta^3(\boldsymbol{p}'-\boldsymbol{p})\,. \tag{2.5}$$

This can be written

$$\delta\varepsilon = \lim_{\Lambda\to\infty}\left\{\frac{e^2}{2}\int\frac{i\,\mathrm{d}^4 q}{(2\pi)^4}\left[\frac{1}{\boldsymbol{q}^2}-\frac{1}{\boldsymbol{q}^2+\Lambda^2}\right]T(q)-\frac{e^2\Lambda}{8\pi}\cdot 2\right\}, \tag{2.6}$$

where $T(q)$ is the time ordered amplitude

$$T(q) := \int \mathrm{d}^4 x\, \mathrm{e}^{iqx}\left(\begin{matrix}\varepsilon_B & j\\ \boldsymbol{p} & s\end{matrix}\middle| -i\mathscr{T}\varrho(x)\,\varrho(0)\middle|\begin{matrix}\varepsilon_B & j\\ \boldsymbol{p} & s\end{matrix}\right) \tag{2.7}$$

and clearly $\varrho(x) = \mathrm{e}^{itH_{\text{st}}}\,\varrho(\boldsymbol{x},0)\,\mathrm{e}^{-itH_{\text{st}}}$.

$T(q)$ can be represented by

$$\begin{aligned} T(q) = F^2(\boldsymbol{q}^2)&\left[\frac{1}{v-\dfrac{\boldsymbol{q}^2}{4m}+i0}-\frac{1}{v+\dfrac{\boldsymbol{q}^2}{4m}-i0}\right]\\ &+\boldsymbol{q}^2\int_{\frac{\boldsymbol{q}^2}{4m}+\varepsilon_B}^{\infty}\frac{\mathrm{d}v'}{m}\,\sigma(v',\boldsymbol{q}^2)\left[\frac{1}{v-v'+i0}-\frac{1}{v+v-i0}\right]\end{aligned} \tag{2.8}$$

with the pole terms and the dispersion integral corresponding to the single particle (elastic) and inelastic contributions respectively:

$T =$ 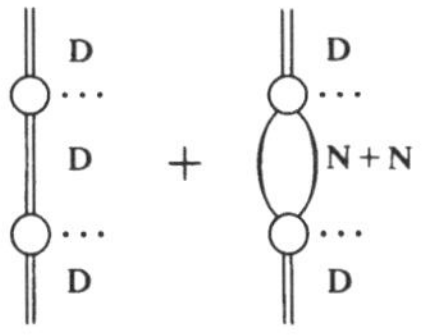

where we defined $v := q^0 - \dfrac{\boldsymbol{pq}}{2m}$

$F(\boldsymbol{q}^2)$ is the elastic deuteron formfactor defined by*

$$\left(\begin{matrix}\varepsilon_B & j\\ \boldsymbol{p}' & s\end{matrix}\middle|\varrho(0)\middle|\begin{matrix}\varepsilon_B & j\\ \boldsymbol{p} & s\end{matrix}\right) = F(\boldsymbol{q}^2)\,,\qquad \boldsymbol{q} := \boldsymbol{p}'-\boldsymbol{p} \tag{2.9}$$

---

* For simplicity we take the spin of the "deuteron" $j=0$ then there is only one formfactor.

it can by represented by the Fourier transform of the square of the bound-state wave function $\varphi_B(x)$

$$F(\boldsymbol{q}^2) = 2 \int \mathrm{d}^3 x \, \mathrm{e}^{i\boldsymbol{q}\frac{\boldsymbol{x}}{2}} \, |\varphi_B(\boldsymbol{x})|^2 \tag{2.10}$$

with the normalisation $\int \mathrm{d}^3 x |\varphi_B(\boldsymbol{x})|^2 = 1$.

From the short distance behaviour $\varphi_B(\boldsymbol{x}) \xrightarrow[r\to 0]{} r^j$ we conclude the asymptotic behaviour

$$F(\boldsymbol{q}^2) \xrightarrow[\boldsymbol{q}^2\to\infty]{} \frac{\text{const}}{(\boldsymbol{q}^2)^{2+j}} \, . \tag{2.11}$$

Analogously would in the case that several single-particle states with spin j' contribute follow

$$\left(\begin{matrix} \varepsilon_B' & j' \\ \boldsymbol{p}' & s \end{matrix} \middle| \varrho(0) \middle| \begin{matrix} \varepsilon_B & j \\ \boldsymbol{p} & s \end{matrix} \right) \xrightarrow[\boldsymbol{q}^2\to\infty]{} \frac{\text{const}}{(\boldsymbol{q}^2)^{2+\frac{1}{2}(j'+j)}} \, . \tag{2.12}$$

Inserting $T(q)$ into the formula for the level shift we obtain in the limit $\Lambda \to \infty$

$$\delta\varepsilon = \frac{e^2}{2} \int \frac{\mathrm{d}^3 q}{(2\pi)^3} \left\{ \frac{F^2(\boldsymbol{q}^2)}{\boldsymbol{q}^2} + \frac{1}{\boldsymbol{q}^2} \left[\boldsymbol{q}^2 \int\limits_{\frac{\boldsymbol{q}^2}{4m}+\varepsilon_B}^{\infty} \frac{\mathrm{d}v}{m} \sigma(v, \boldsymbol{q}^2) - 2 \right] \right\} . \tag{2.13}$$

We see from this formula that the single particle contributions are always positive and convergent whereas for the inelastic part we have the sum rule

$$F(0) + F(\boldsymbol{q}^2) = F^2(\boldsymbol{q}^2) + \boldsymbol{q}^2 \int\limits_{\frac{\boldsymbol{q}^2}{4m}+\varepsilon_B}^{\infty} \frac{\mathrm{d}v}{m} \sigma(v, \boldsymbol{q}^2) , \tag{2.14}$$

which guarantees the finiteness of the level shift $\delta\varepsilon$. From the sum rule we conclude the asymptotic behaviour

$$\lim_{\boldsymbol{q}^2\to\infty} \boldsymbol{q}^2 \int\limits_{\frac{\boldsymbol{q}^2}{4m}+\varepsilon_B}^{\infty} \frac{\mathrm{d}v}{m} \sigma(v, \boldsymbol{q}^2) = 2 \, . \tag{2.15}$$

Since $\sigma(v, \boldsymbol{q}^2)$ is connected with the inelastic electron-deuteron cross section 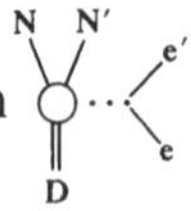

we get the large momentum-transfer behaviour

$$\lim_{q^2\to\infty} \left((\boldsymbol{q}^2)^2 \, E_{\text{Lab}} \frac{\mathrm{d}\sigma^{\text{inel.}}}{\mathrm{d}\boldsymbol{q}^2} \right) = 4\pi m_e \alpha^2 \tag{2.16}$$

i.e. the inelastic cross section behaves for large momentum transfer like the elastic cross section of a point particle which is

$$(\boldsymbol{q}^2)^2 E_{\text{Lab}} \frac{\mathrm{d}\sigma^{\text{el}}}{\mathrm{d}\boldsymbol{q}^2} = \text{const}\,. \tag{2.17}$$

Using the sum rule (2.14) we finally can write the level shift

$$\delta\varepsilon = \frac{e^2}{2} \int \frac{\mathrm{d}^3 q}{(2\pi)^3} \frac{1}{\boldsymbol{q}^2} F(\boldsymbol{q}^2)\,, \tag{2.18}$$

which is the form one usually writes it in nonrelativistic theory. This form, however, has probably no generalisation to the more complicated case of a relativistic theory.

3. The Electromagnetic Mass Shift of Hadrons

a) The Cottingham [3] Formula

The electromagnetic mass shift of a hadron is to order e^2 given by the formula

$$2M\,\delta M = \frac{e^2}{2} \int \frac{i\,\mathrm{d}^4 q}{(2\pi)^4} \frac{g^{\mu\nu}}{q^2 + i0} T_{\mu\nu}(q)\,. \tag{3a.1}$$

Where $T_{\mu\nu}(q)$ in suitable cases is the truncated time ordered amplitude

$$T_{\mu\nu}(q) := \int \mathrm{d}^4 x\, \mathrm{e}^{iqx} \frac{1}{2j+1} \sum_s \langle ps| - i\mathscr{T} j_\mu(x) j_\nu(0)\, |ps\rangle_{\text{tr.}} \tag{3a.2}$$

in the covariant normalisation

$$\langle p' s' \,|ps\rangle = (2\pi)^3\, 2p^0 \delta^3(\boldsymbol{p} - \boldsymbol{p}')\, \delta_{ss'}\,, \tag{3a.3}$$

which then is a covariant and gauge invariant quantity and therefore can be written

$$T_{\mu\nu}(q) = (q_\mu q_\nu - q^2 g_{\mu\nu})\, T_1(v, q^2)$$
$$\left\{ v^2 g_{\mu\nu} + \frac{q^2}{M^2} p_\mu p_\nu - \frac{v}{M^2} (p_\mu q_\nu + p_\nu q_\mu) \right\} T_2(v, q^2)\,, \tag{3a.4}$$

where now we defined $v := \dfrac{pq}{M}$.

The trace T^μ_μ which enters into the mass shift can be split into a Coulomb-part, and a Lamb-part coming from the 2 transverse degrees of freedom of the electromagnetic field:

$$T^{\mu}_{\mu}(q) = \underbrace{q^2(T_2 - T_1)}_{\text{Coulomb}} + 2\underbrace{(v^2 T_2 - q^2 T_1)}_{\text{Lamb}} . \tag{3a.5}$$

Following *Cottingham* [3] we rotate the q^0-integration in (3a.1) to the imaginary axis which is possible due to the location of the singularities of the time ordered amplitude and the high q^0-behaviour $\lim\limits_{q^0 \to \infty} q^0 T^{\mu}_{\mu} = 0$ which follows from the vanishing equal-time-commutator of the electromagnetic current j_μ. Introducing instead of v and q^2 new variables $z, \varkappa^2$ by the relations

$$\varkappa^2 := -q^2, \qquad \varkappa^2 z^2 := -v^2$$

we can (3a.1) bring to the form

$$2M\,\delta M = \frac{\alpha}{4\pi} \int\limits_0^\infty \mathrm{d}\varkappa^2 \varkappa^2 \cdot \frac{1}{\pi} \int\limits_{-1}^{1} \mathrm{d}z \sqrt{1-z^2}\, \{(T_2 - T_1) + 2(z^2 T_2 - T_1)\} . \tag{3a.6}$$

We see that only spacelike momentum transfers q^2 enter the formula and for such q^2 we can write dispersion relations for T_1, T_2 (assuming for the moment unsubtracted ones)

$$T_k = \text{pole term} + \frac{1}{\pi} \int\limits_{v_0}^{\infty} \frac{\mathrm{d}v'^2 t_k(v', q^2)}{v^2 - v'^2}, \qquad v_0 = -\frac{q^2}{2M} + m\left(1 + \frac{m}{2M}\right). \tag{3a.7}$$

Again the pole terms are related to the elastic form factors and the t_k to the inelastic electroproduction cross-sections:

$$t_2 - t_1 \sim \frac{\mathrm{d}\sigma^{\text{inel}}_{\text{coul}}}{\mathrm{d}q^2\,\mathrm{d}v}, \qquad \frac{v^2}{q^2} t_2 - t_1 \sim \frac{\mathrm{d}\sigma^{\text{inel}}_{\text{trans}}}{\mathrm{d}q^2\,\mathrm{d}v} . \tag{3a.8}$$

Thus finally the electromagnetic mass differences have been expressed by measurable quantities. Clearly in the case of subtracted dispersion relations additional subtraction constants, which are in fact functions of q^2, enter. Let me first discuss the elastic contributions.

b) The Single Particle (Elastic) Contribution to the Electromagnetic Mass-Shift

We may restrict ourselves to spin 1/2 particles since the spin 0 case is obtained by leaving out the magnetic part i.e. by $F_M \equiv 0$ in the following formulae. Then we have the expressions*

$$T_1^{\text{S.P.}} = \frac{2q^2}{q^4 - 4M^2 v^2}(F_E^2 - F_M^2), \tag{3b.1}$$

$$T_2^{\text{S.P.}} = \frac{8M^2}{q^4 - 4M^2 v^2}\left(F_E^2 - \frac{q^2}{4M^2} F_M^2\right) \tag{3b.2}$$

and from these

$$-g^{\mu\nu} T_{\mu\nu}^{\text{S.P.}} = 2F_E^2 \frac{3q^4 - 4M^2 q^2 - 8M^2 v^2}{q^4 - 4M^2 v^2} - 4F_M^2 \frac{q^4 - q^2 v^2}{q^4 - 4M^2 v^2}. \tag{3b.3}$$

This finally gives the mass shift

$$2M\,\delta M^{\text{S.P.}} = \frac{\alpha}{2\pi}\int_0^\infty d\varkappa^2 \left\{ F_E^2(\varkappa^2) \cdot \frac{1}{\pi}\int_{-1}^{1} dz \sqrt{1-z^2}\, \frac{\overbrace{(4M^2+\varkappa^2)}^{\text{Coulomb}} + \overbrace{2(\varkappa^2 + 4M^2 z^2)}^{\text{transverse}}}{\varkappa^2 + 4M^2 z^2} - 2F_M^2 \cdot \varkappa^2 \cdot \frac{1}{\pi}\int_{-1}^{1} dz (1-z^2)^{3/2} \cdot \frac{1}{\varkappa^2 + 4M^2 z^2} \right\}. \tag{3b.4}$$

From this formula we see that the electric contribution to the mass-shift is positive, as it should, since like pieces of charge repell. This part is the sum of a Coulomb-shift and a "Lamb"-shift comming from the two transverse degrees of freedom of the radiation field. The Coulomb-shift is larger than the "Lamb"-shift due to only *one* transverse degree of freedom. Furthermore the magnetic contribution to the mass-shift is negative, also as we expect, since parallel currents attract.

Performing the z-intergration we obtain

$$2M\,\delta M^{\text{S.P.}} = \frac{\alpha}{2\pi}\int_0^\infty d\varkappa^2 \left\{ F_E^2(\varkappa^2)\Bigg[\underbrace{\left(1+\frac{\varkappa^2}{4M^2}\right)\left(\sqrt{1+\frac{4M^2}{\varkappa^2}}-1\right)}_{K_{\text{coul}}} + \underbrace{1}_{K_{E,\text{trans}}} \Bigg] - F_M^2(\varkappa^2) \cdot \underbrace{\frac{\varkappa^2}{2M^2}\left[\left(1+\frac{\varkappa^2}{4M^2}\right)\left(\sqrt{1+\frac{4M^2}{\varkappa^2}}-1\right)-\frac{1}{2}\right]}_{K_M} \right\}. \tag{3b.5}$$

---

* For spin 0 particles F_E^2 is the square of the usual formfactor F normalised to $F(0)=1$. For spin 1/2 particles the connection with the Sachs-formfactors $G_{E,M}$ is

$$G_{E,M}^2 = \left(1 - \frac{q^2}{4M^2}\right) F_{E,M}^2 .$$

Looking at Fig. 1 where these Kernels are plotted together with the formfactors we see, that the elastic contribution to the proton-neutron mass differences always will have the wrong sign; the same will happen in the K^+, K^0 case.

On the contrary the $\pi^+\pi^0$ mass difference is given to a good approximation by the elastic contribution. We conclude that the $I=1$ mass differences must essentially be determined by the inelastic contributions on which we so far have little data. Moreover it seems that even if we had the data the integral over the inelastic cross section entering the mass formula will diverge, due to the slow decrease of the total inelastic cross section for large q^2. This disturbing feature follows from arguments given by *Bjorken* [1] which I shall discuss in the next section.

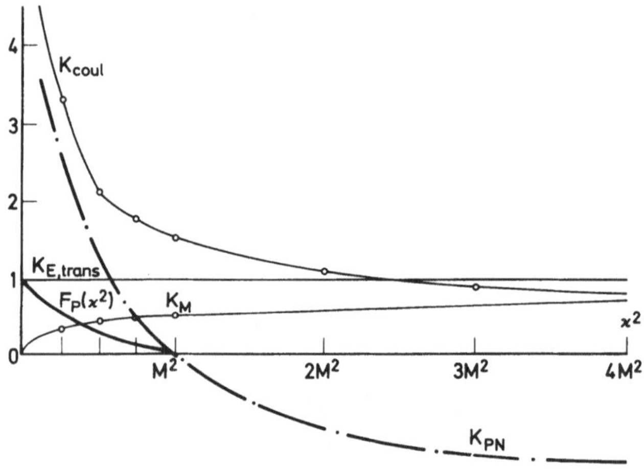

Fig. 1. The various Kernels of Eq. (3b.5) are plotted versus $\varkappa^2$. K_{PN} is the Kernel relevant for the Proton-Neutron mass difference, $F_P(\varkappa^2)$ the Proton formfactor

4. The Connection Between the Asymptotic Behaviour of the "Compton"-Amplitude and Equal Time Commutators of the Electromagnetic Current

For the problem of mass differences we can disregard the $I=0$ part in the $\mathscr{T}$-Produkt and therefore also the truncation in (3a.2). Let us discuss now what the consequences of various equal time commutators on $T_{\mu\nu}$ are:

The explicit form of the various components of $T_{\mu\nu}$ according to (3a.4) is

$$
\begin{aligned}
T^{00} &= \boldsymbol{q}^2 T_1 - \left[\boldsymbol{q}^2 \frac{p^{02}}{M^2} - \frac{(\boldsymbol{pq})^2}{M^2}\right] T_2 , \\
T^{0k} &= q^0 \left\{ q^k T_1 + \left(q^k \frac{p^{02}}{M^2} - p^k \frac{\boldsymbol{pq}}{M^2}\right) T_2 \right\} \\
&\quad + \frac{p^0}{M^2} [q^k(\boldsymbol{pq}) - p^k \boldsymbol{q}^2]\, T_2 , \\
T^{kl} &= q^{02} \left\{ \delta^{kl} \left(T_1 - \frac{p^{02}}{M^2} T_2 \right) + \frac{p^k p^l}{M^2} T_2 \right\} \\
&\quad + q^0 \frac{p^0}{M^2} [2\delta^{kl} \boldsymbol{pq} - (p^k q^l + p^l q^k)]\, T_2 \\
&\quad + \text{terms constant in } q^0 \text{ times } T_k .
\end{aligned}
\tag{4.1}
$$

On the other hand we have the asymptotic connection ($I \neq 0$)

$$
\begin{aligned}
T^{\mu\nu} \xrightarrow[\substack{q^0 \to \infty \\ \boldsymbol{q}\,\text{fixed}}]{} & \frac{1}{q^0} \int d^3x\, e^{-i\boldsymbol{qx}} \langle \boldsymbol{p} | [j^\mu(\boldsymbol{x}, 0), j^\nu(0)] | \boldsymbol{p} \rangle \\
& + \frac{1}{q^{02}} \int d^3x\, e^{-i\boldsymbol{qx}} \langle p | i[\dot{j}^\mu(x, 0), j^\nu(0)] | p \rangle + \cdots
\end{aligned}
\tag{4.2}
$$

$$
\begin{aligned}
&\text{From} \quad [j^0, j^0]_{ET} = 0 \quad \text{we conclude} \quad q^0 T_{1,2} \xrightarrow[q^0 \to \infty]{} 0 , \\
&\text{From} \quad [j^0, j^k]_{ET} = 0 \quad \text{we conclude} \quad q^{02} T_{1,2} \longrightarrow 0 , \\
&\text{From} \quad [j^k, j^l]_{ET} = 0 \quad \text{we conclude} \quad q^{03} T_{1,2} \longrightarrow 0 .
\end{aligned}
\tag{4.3}
$$

Furthermore if $[\dot{j}^k(\boldsymbol{x}, 0), j^l(0)] \neq 0$ it follows that

$$
q^{04} T_{1,2} \to \text{const} \neq 0 . \tag{4.4}
$$

This *in general* leads to logarithmic divergent mass differences by the following argument:

For the calculation of the mass differences which we can do in the rest frame $\boldsymbol{p} = 0$ we need the "Compton"-amplitude (3a.2) everywhere in the $(q^0, |\boldsymbol{q}|) = (i\lambda, |\boldsymbol{q}|)$ plane (λ = real), whereas from the equal time comuntator we sofar have information on it only for $\lambda \to \infty$, $|\boldsymbol{q}|$ fixed. There is no rigorous way to extend this information to obtain the asymptotic behaviour in any direction of the $(\lambda, |\boldsymbol{q}|)$ plane. *Bjorken* [1] has, however, given arguments what *at best* can be expected. This arguments depend on the analytic properties of the "Compton"-amplitude which for $q^2 \leq 0$

are described by the dispersion relations

$$T_k(v, q^2) = \text{pole terms} + \frac{1}{\pi}\int_{v_0^2}^{\infty} \frac{dv'^2\, t_k(v', q^2)}{v^2 - v'^2}$$
$$v_0 = -\frac{q^2}{2M} + \text{const} \tag{4.5}$$

Bjorken [1] observes now that the lower boundary of the integral v_0^2 recedes to ∞ for $\lambda \to \infty$ as fast as λ^4, therefore the denominator $v'^2 - v^2 = v'^2\left(1 + \frac{\lambda^2}{v'^2}\right) \to v'^2\left(1 + \frac{4M^2}{\lambda^2}\right)$ can be approximated by v'^2 with an error less than ε^2 in the whole $(\lambda, |\boldsymbol{q}|)$ region defined by

$$\left(\lambda \pm \frac{M}{\varepsilon}\right)^2 + \boldsymbol{q}^2 > \left(\frac{M}{\varepsilon}\right)^2, \tag{4.6}$$

i.e. outside of the two shaded circles of Fig. 2.

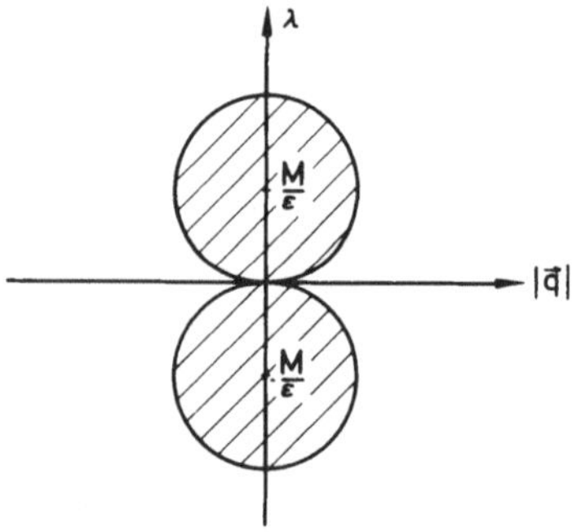

Fig. 2. Outside the two circles the amplitudes $T_k(\lambda, |\boldsymbol{q}|)$ depend on the direction in which one goes to infinity according to Eqs. (4.6a) and (4.8) respectively

Outside of the two circles we therefore have (omitting the pole terms) for T_k the asymptotic form

$$T_k^{\text{asympt.}} = -\frac{1}{\pi}\int_{v_0^2}^{\infty} \frac{dv'^2}{v'^2}\, t_k(v', q^2) = f_k(q^2). \tag{4.6a}$$

The assumption implied here is of course that for $\lambda \to \infty$ the integrand $t_k(v', \lambda^2)$ does not blow up.

Thus we see that in the case of unsubtracted dispersion relations for T_k the asymptotic behaviour is the same in all directions of the $(\lambda, |\boldsymbol{q}|)$

plane. For a subtraced dispersion relation

$$T_k = a_k(q^2) + \frac{v^2}{\pi} \int_{v_0^2}^{\infty} \frac{dv'^2}{v'^2} \frac{t_k(v', q^2)}{v^2 - v'^2} \tag{4.7}$$

we conclude the same way

$$T_k^{\text{asympt.}} = a_k(q^2) + v^2 g_k(q^2) \tag{4.8}$$

outside the two circles of Fig. 2.

Concentrating now on the commutators of the field algebra model [2] we find that we have to use a subtracted form for T_1* and find the asymptotic expressions

$$T_1^{\text{asympt.}} = \frac{M^2}{(q^2)^2} \left\{ A + \frac{v^2}{q^2} B \right\}, \qquad T_2^{\text{asympt.}} = \frac{M^2}{(q^2)^2} B, \tag{4.9}$$

with the constants A and B given by

$$\frac{1}{2s+1} \sum_s \sum_{\alpha=1,2,4,5} \langle ps | j_\mu^\alpha(0)\, j_\nu^\beta(0) + a_\mu^\alpha(0)\, a_\nu^\beta(0) | ps \rangle = \left(\frac{m_\varrho}{g_\varrho} \right)^2 \left\{ -g_{\mu\nu} A + \frac{p_\mu p_\nu}{M^2} B \right\} M^2 . \tag{4.10}$$

This leads to the divergent parts of the mass shifts

$$\delta M^{\text{div.}} = \frac{3}{8} \alpha \left(\frac{g_\varrho^2}{4\pi} \right) \left(\frac{4A - B}{m_\varrho^2} \right) \int_0^\infty \frac{d\varkappa^2}{\varkappa^2}, \tag{4.11}$$

Whenever $4A - B \neq 0$ the mass difference will be infinite to order e^2.

Thus we have the alternatives:

1. Use field theoretic models where the commutator $[\dot{j}_\mu(\boldsymbol{x}, 0), j_\nu(0)]$ = universal so that the divergence comming from it can be removed by a universal mass renormalisation; mass differences to order e^2 would then the finite.

2. Mass differences are infinite to order e^2 due to the unallowed use of the perturbation expansion. This would happen if the exact mass differences had e.g. the form

$$\delta M = \alpha [R(\alpha) \ln \alpha + S(\alpha)] \tag{4.12}$$

with $R(0)$, $S(0)$ finite.

Such a behaviour might also hold in the case of radiative corrections to weak decays.

---

* *Harari* [4] has given arguments that for the $I = 1$ part of T_1 a subtraction is necessary because of the A_2-Regge trajectorie contributing to it.

To order e^2 then one would have to introduce a cut off Λ, which merely signals the unallowed expansion of say $\ln \alpha$ around $\alpha = 0$. Summing the series we would expect terms of the type $\ln\left(\alpha + \frac{m}{\Lambda}\right)$ in which the limit $\Lambda \to \infty$ can be readily performed.

3. The theories we have sofar are all incomplete at high energies, some effective cut off has to be used to get finite results which are then hoped to be relevant.

5. On a Possible Connection of the $I=1$ Electromagnetic Mass Differences with the Nonleptonic Weak Decays

The possibility that the divergence of the electromagnetic mass differences to order e^2 might be due to the unallowed use of a power expansion in e for these quantities opens an interesting connection with the nonleptonic weak decays within the model of field algebra [2], where the relevant equal time commutator has the form of the local current $\times$ current nonleptonic weak interaction $\mathscr{H}_{PC}$. More precisely it can be related to $\mathscr{H}_{PC}$ by SU_3 transformations. From the experimentally observed octet dominance of the one particle matrixelements of $\mathscr{H}_{PC}$ we conclude that only the octet part of the electromagnetic mass differences, i.e. not their $I=2$ part, is mainly determined by the "divergent" part. Furthermore we have the result that the "divergent" part of the mass differences has the same d/f ratio as the corresponding weak amplitudes, which is true to a good approximation for the actual mass differences as we heard in the lectures by Prof. *Stech*. This suggests that the "divergent" part gives in fact the bulk of the actual $I=1$ mass differences. This can be checked by calculating the ratio

$$\left|\frac{\Delta^{\text{div}} m_{K^+K^0}}{\Delta^{\text{div}} m_{\text{NP}}}\right| = 2.7 \qquad (= 3 \text{ experimentally}),$$

which is cut off independent and well in agreement with experiment. Furthermore fitting the neutron-proton mass difference one obtains for the cut off a value $m_\varrho/\Lambda \approx 2\alpha$ in accordance with the philosophy about the origin of Λ. For the detailes I refer you to the paper by *V. Müller* and myself [5]. There it is also estimated that the contribution of the "divergent" part to the $\pi^+\pi^0$ mass difference, which is $I=2$, is certainly less than 1 MeV, probably even smaller. This would explain why essentially all calculations of this mass difference have been successful, especially also the soft pion calculation of *Das* [6] et al. which was discussed in Prof. Wick's lecture.

References

1. *Bjorken, J. D.:* Phys. Rev. **148**, 1467 (1966).
2. *Lee, T. D., S. Weinberg,* and *B. Zumino:* Phys. Rev. Lett. **18**, 1029 (1967).
3. *Cottingham, W. N.:* Ann. Phys. (N. Y.) **25**, 424 (1963).
4. *Harari, H.:* Phys. Rev. Lett. **17**, 1303 (1966).
5. *Müller, V.*, and *J. Rothleitner:* Nucl. Phys. B **5**, 373 (1968).
6. *Das, T., G. S. Guralnik, V. S. Mathur, F. E. Low,* and *J. E. Young:* Phys. Rev. Letters **18**, 759 (1967).

Univ.-Prof. Dr. *J. Rothleitner*
Institut für Theoretische Physik II
der Universität Innsbruck
A-6020 Innsbruck

Nonleptonic Decays and Mass Differences of Hadrons

B. Stech

Contents

1. Introduction

Nonleptonic weak interactions [1, 2] are not very well understood. The reason is that the Hamiltonian responsible for the interaction is the product of two hadron currents at small or vanishing space-time distances. Such operator products are difficult to handle. Fortunately, however, there is one concept which can be applied just because of the short range of the current times current product: *In the point interaction limit simple commutation relations of the $SU\,3 \times SU\,3$ charges with the weak Hamiltonian hold* [2, 3]. In the first part of this lecture we will extensively use the chiral symmetry of the Hamiltonian obtained in this way. At this occasion we will learn that the "*soft π-meson limit*" *is nothing else but an approximation to a very simple sum rule*, which can immediately be obtained from the commutation relation [4, 2, 5]. The corrections to the limit vanish only in a fictitious world with $m_\pi = 0$ and a conserved axial vector current. These corrections can be exhibited and discussed. An important part of it has to be taken care of by the use of an additional commutation relation. We will also see that the *corrections to the soft pion approximation are smallest for a kinematical configuration in which the particle states in question have vanishing space momenta* [4]. In general, this is not the kinematical configuration of a decay process. However, *usual on-mass shell dispersion relations can be used to relate the different kinematical points.*

In the second part of this lecture we will discuss the connection between S-wave decay amplitudes and the mass differences of hadrons, especially electromagnetic mass-differences. The connection arises because the Hamiltonian responsible for electromagnetic mass differences is again of the current × current form.

2. Nonleptonic Decays

In view of the success of the current × current type Hamiltonian in describing leptonic and semi-leptonic decays, one assumes this structure to be significant also for nonleptonic weak decays. Thus the product of the hadron current with itself should form the Hamiltonian responsible for these decays.

$$\mathscr{H}(x)=\frac{G}{\sqrt{2}}J_\lambda^+(x)\,J^\lambda(x)\,, \qquad (1)$$
$$J_\lambda^{\text{hadron}}(x)=\cos\theta(V_\lambda^{1+i2}(x)-A_\lambda^{1+i2}(x))+\sin\theta(V_\lambda^{4+i5}(x)-A_\lambda^{4+i5}(x))\,,$$

θ is Cabibbo's angle. $J_\lambda^{\text{hadron}}$ is the current of strong interaction belonging to the (8,1) representation of the chiral $SU\,3\times SU\,3$ group. The corresponding equal-time commutation relations with the $SU\,3\times SU\,3$ charges $F^\alpha(t)$ and $F_5^\alpha(t)$ are supposed to be strict to all orders of strong interaction. The current obeys in particular the chiral symmetry property

$$\begin{gathered}[F_5^\alpha(t),J_\lambda^{\text{hadron}}(x)]_{x_0=t}=[F^\alpha(t),J_\lambda^{\text{hadron}}(x)]_{x_0=t}\,,\\ F^\alpha(t)=\int V_0^\alpha(x)\,\mathrm{d}^3x\,,\qquad F_5^\alpha(t)=-\int A_0^\alpha(x)\,\mathrm{d}^3x\,.\end{gathered} \qquad (2)$$

A typical decay is the process $\Lambda\to p+\pi^-$. Only strangeness changing transitions can be observed. In strangeness conserving processes the strong interaction dominates by many orders of magnitude. The effects of parity non-conservation due to the strangeness conserving nonleptonic weak interaction have so far been detected only in very precise measurements of radiative transitions in nuclei [6].

The strangeness changing decays arise from the product of the isospin vector current with the isospin spinor current. The transition amplitude T_{fi} is given by [7]

$$T_{fi}/(2\pi)^4=\langle f|\,\mathscr{H}_{\text{weak}}^{\text{nonleptonic}}\,|i\rangle\,,$$
$$T_{fi}/(2\pi)^4=\frac{G}{\sqrt{2}}\sin\theta\cos\theta\frac{1}{2}\langle f|\,\{(V_\lambda(0)-A_\lambda(0))^{1-i2},(V^\lambda(0)-A^\lambda(0))^{4+i5}\}_+ + \text{h.c.}\,|i\rangle\,. \qquad (3)$$

A symmetrized form with respect to the $SU\,3$ indices has been taken. For the moment we will not question the local form of the current product although an operator product at equal space-time points is not likely to exist.

The symmetrized product of the two octet current operators can be decomposed into an octet and a "27" part. In discussing $SU\,3$ or $SU\,3 \times SU\,3$ transformation properties we should keep in mind that the currents are – according to their commutation relations – supposed to be strict octets irrespective of the amount of symmetry breaking in the particle states. Thus the decomposition of the current product into octet and "27" is well defined at least as long as the interaction is local.

Experimentally, one observes what is called the empirical $|\vec{\Delta I}| = 1/2$ rule for nonleptonic decays: $\mathscr{H}_{\text{weak}}^{\text{nonlept.}}$ transforms as an $I = 1/2$ operator. The $|\vec{\Delta I}| = 3/2$ transition $K^+ \to \pi^+ + \pi^\circ$ has a lifetime about 500 times longer than the corresponding decay $K^\circ \to \pi^+ + \pi^-$ which is allowed by the rule. Thus, the rule holds to about 5% in amplitude. A typical prediction of the $|\vec{\Delta I}| = 1/2$ rule is $\Gamma(\Lambda \to p\pi^-)/\Gamma(\Lambda \to n\pi^\circ) = 2$ which is very well satisfied experimentally (1.97 ± 0.08 for amplitudes squared).

Consequently, the "27" part of the current product does not contribute significantly, at least not its $I = 3/2$ component. Two possible mechanisms have been suggested: i) a dynamical enhancement of the octet part and ii) the addition of neutral currents to the Hamiltonian in such a way that a pure octet results. The latter hypothesis is not so appealing since neutral currents coupled to neutral lepton currents have not been observed. The hypothesis of a dynamical octet dominance has some support from the observation that the self-product of the electromagnetic current shows dynamical octet enhancement as can be seen from the electromagnetic mass splittings in isomultiplets of hadrons.

It follows directly from the $SU\,3$-structure of the currents occuring in Eq. (3) that the octet part of the interaction is the 6th component of an octet. In fact, in the $SU\,3 \times SU\,3$ classification the octet part is the 6th component of the first $SU\,3$ in the (8,1) representation.

The chiral symmetry of the currents (2) has an important consequence for the nonleptonic weak Hamiltonian. If we decompose $\mathscr{H}_{\text{weak}}$ into a parity conserving part $\mathscr{H}^{\text{p.c.}}$ and a parity violating part $\mathscr{H}^{\text{p.v.}}$ we have from Eqs. (1) and (2)

$$\begin{aligned} [F_5^\alpha(t), H^{\text{p.v.}}(t)] &= [F^\alpha(t), H^{\text{p.c.}}(t)]\,, \\ [F_5^\alpha(t), H^{\text{p.c.}}(t)] &= [F^\alpha(t), H^{\text{p.v.}}(t)]\,, \end{aligned} \qquad H(t) = \int d^3x\, \mathscr{H}(\mathbf{x}, t)\,. \tag{4}$$

These commutation relations can be used advantageously to express matrix elements involving π-mesons by simpler forms. The direct use of matrix elements of the commutation relations like (4) is the simplest as well as the best method available for this purpose [2, 4]. As already mentioned in the introduction it gives the soft π-meson results including the corrections.

i) Hyperon Decays

There are seven known hyperon decays (the common shorthand notation for each is given in paranthesis):

$$\begin{aligned}
\Lambda &\to p + \pi^- && (\Lambda^\circ_-)\,, \\
\Lambda &\to n + \pi^\circ && (\Lambda^\circ_\circ)\,, \\
\Sigma^+ &\to p + \pi^\circ && (\Sigma^+_\circ)\,, \\
\Sigma^+ &\to n + \pi^+ && (\Sigma^+_+)\,, \\
\Sigma^- &\to n + \pi^- && (\Sigma^-_-)\,, \\
\Xi^- &\to \Lambda + \pi^- && (\Xi^-_-)\,, \\
\Xi^\circ &\to \Lambda + \pi^\circ && (\Xi^\circ_\circ)\,.
\end{aligned}$$

The final state contains both S and P-waves since parity is not conserved in the transition. The transition amplitude may be written

$$T_{fi}/(2\pi)^4 = \langle B_f \pi | \mathscr{H}^{(0)}_{\text{weak}} | B_i \rangle = \bar{u}_{p_f}(A + B\gamma_5)\, u_{p_i}\,. \tag{5}$$

The quantity A describes the parity violating S-wave decay amplitude S, and B is proportional to the parity conserving P-wave decay amplitude P.

$$A = S\,, \qquad kB = P\,,$$

where (6)

$$k = \sqrt{\frac{(M_i - M_f)^2 - m_\pi^2}{(M_i + M_f)^2 - m_\pi^2}}\,.$$

The decay constant Γ can be written

$$\Gamma = C(|S|^2 + |P|^2) \tag{7}$$

with $C = \dfrac{P_f^c}{4M_i^2}\,((M_i + M_f)^2 - m_\pi^2)\,(2\pi)^8$,

$$p_f^c = |\boldsymbol{p}_f| \quad \text{in C.M. system}\,.$$

The decay parameters α, β, γ are defined as follows

$$\begin{aligned}
\alpha &= \frac{2\,\mathrm{Re}(S^* P)}{|S|^2 + |P|^2}\,, \\
\beta &= \frac{2\,\mathrm{Im}(S^* P)}{|S|^2 + |P|^2}\,, \\
\gamma &= \frac{|S|^2 - |P|^2}{|S|^2 + |P|^2}\,.
\end{aligned} \tag{8}$$

The parameter α is determined by the angular distribution of the decay products relative to the hyperon spin. The parameters β and γ can be determined by measurements of initial and final baryon spins.

The final state interaction introduces only small phase factors in the partial waves as is known from π-nucleon scattering at small energy. If in addition CP-violating effects are small as observed in K°-decays β has to be small in agreement with the data.

From the measured values of Γ, α, and the sign of γ the amplitudes S and P can be found. They are uniquely determined up to a common sign for each decay channel. The parameter γ is not known for the decays $\Sigma_\circ^+$, $\Lambda_\circ^\circ$, Ξ_-^-, $\Xi_\circ^\circ$. In these cases P and S can be interchanged. Table 1 summarizes the result. Arbitrarily $P(\Sigma_\circ^+)$ has been chosen negativ. The other open signs are chosen by requiring fits to the $|\vec{\Delta I}|=1/2$, rule

$$\begin{aligned} -\Sigma_-^- + \Sigma_+^+ &= \sqrt{2}\,\Sigma_\circ^+ , \\ \Lambda_-^\circ &= -\sqrt{2}\,\Lambda_\circ^\circ , \\ \Xi_-^- &= -\sqrt{2}\,\Xi_\circ^\circ \end{aligned} \tag{9}$$

and the Lee-Sugawara rule

$$2\Xi_-^- + \Lambda_-^\circ = \sqrt{3}\,\Sigma_\circ^+ \tag{10}$$

for both quantities A and B. The latter relation has theoretical support from octet dominance as we will see. Of course, the signs in the above relations depend on $SU\,3$ conventions. We used the ones chosen by *Gell-Mann*.

Table 1

	Λ_-°	$\Lambda_\circ^\circ$	Σ_-^-	$\Sigma_\circ^+$	Σ_+^+	Ξ_-^-	$\Xi_\circ^\circ$
k	0.0533	0.0528	0.101	0.0998	0.0975	0.0620	0.0607
$(2\pi)^{9/2}\,A\cdot 10^7$	-3.36	2.36	-4.04	3.40	-0.04	4.42	-3.47
$(2\pi)^{9/2}\,B\cdot 10^7$	-23.3	18.2	$+0.8$	-25.1	-41.4	-13.4	9.7

For a theoretical discussion it is necessary to describe the amplitudes A and B as functions of *independent* variables s, t and u defined by

$$s=(p_f+q)^2, \quad t=(p_f-p_i)^2, \quad u=(p_i-q)^2 . \tag{11}$$

The momentum of the π-meson is denoted by q. Clearly, in the final expression we have to put

$$s=M_i^2, \quad t=m_\pi^2, \quad u=M_f^2 .$$

We write

$$\langle B_f \pi | \mathscr{H}^{\text{p.v.}} | B_i \rangle = (\bar{u}_{p_f} u_{p_i})\, f(s,t,u) + (\bar{u}_{p_f} \gamma q u_{p_i})\, g(s,t,u)$$

and (12)

$$\langle B_f \pi | \mathscr{H}^{\text{p.c.}} | B_i \rangle = (\bar{u}_{p_f} \gamma_5 u_{p_i})\, h(s,t,u) + (\bar{u}_{p_f} \gamma_5 \gamma q u_{p_i}) j(s,t,u)$$

and define

$$\begin{aligned} \langle B' | \mathscr{H}^{\text{p.c.}} | B \rangle &= (\bar{u}_{p'} u_p)\, a(t)\,, \\ \langle B' | \mathscr{H}^{\text{p.v.}} | B \rangle &= (\bar{u}_{p'} \gamma_5 u_p)\, b(t)\,. \end{aligned} \tag{13}$$

In the limit of strict $SU\,3$ invariance and with CP-invariance $b(t)$ is zero. This is seen by looking upon the crossed matrix element: The $SU\,3$ indices of baryon and antibaryon occur in an antisymmetric form. Thus, they do not couple to the symmetric U-spin 1 combination occuring in $\mathscr{H}^{\text{p.v.}}$. However, by this argument we can not expect $b(t)$ to be smaller than $a(t)$. On the contrary both quantities are connected by $SU\,3 \times SU\,3$ and should fall off for large t roughly in the same way. At low values of t the K-meson pole even enhances $b(t)$ compared to $a(t)$. As a consequence we except baryon pole diagramms to contribute to both, the S- and P-wave decays.

In spite of $SU\,3$ symmetry breaking we can use $SU\,3$ to relate the different baryon states occuring in Eq. (13) with each other leaving the $SU\,3$ component of $\mathscr{H}$ unchanged, of course. This is compatible with and suggested by a dispersion theoretical treatment. The quantity $b(t)$ is expected to have a dominant contribution from a K_1°-particle intermediate state. The matrix element $\langle 0 | \mathscr{H}^{\text{p.v.}} | K_1^\circ \rangle$ breaks the symmetry (K_1° is the 7th component of an octet, the octet part of $\mathscr{H}^{\text{p.v.}}$ is a 6th component). But from the coupling of the K_1° particle to the baryons one has octet transformation properties with a D/F-ratio of $D/F \cong 3/2$. Thus, $SU\,3$ relations can still be used to a certain extent.

We choose

$$b_{ji}(t) = (D D_{ji} + F F_{ji})\, b(t) \quad \text{with} \quad D + F = 1\,, \tag{14}$$

where D_{ji} and F_{ji} are $SU\,3$ coefficients. For the quantity $a_{ji}(t)$ we use a similar ansatz

$$a_{ji}(t) = (d D_{ji} + f F_{ji})\, a(t) \quad \text{with} \quad d + f = 1\,.$$

If a scalar object dominates the dispersion relation for $a(t)$ the same object will also determine the $SU\,3$ mass-splitting. The $SU\,3$ mass-splittings transform as an octet and have a d/f ratio $d/f \cong -\frac{1}{3}$.

In the following we will find it necessary to use a modified form of $\mathscr{H}^{\text{p.v.}}$ from which the K_1°-part is removed. We define

$$\hat{\mathscr{H}}^{\text{p.v.}} = \mathscr{H}^{\text{p.v.}} - r\, \partial^\lambda A_\lambda^7\,, \tag{15}$$

where the constant r is determined by the formula

$$\langle 0|\hat{\mathscr{H}}^{\text{p.v.}}|K_1^\circ\rangle = 0\,.$$

The one-particle matrix elements of $\hat{\mathscr{H}}^{\text{p.v.}}$ are expected to be an order of magnitude smaller than those of $\mathscr{H}^{\text{p.v.}}$. On the other hand, a transition matrix element of $\hat{\mathscr{H}}^{\text{p.v.}}$ equals the one of $\mathscr{H}^{\text{p.v.}}$ at the point where energy and momentum are conserved. Unfortunately, the constant r cannot be calculated without having more information about $\mathscr{H}^{\text{p.v.}}$ than we have. The reason is again that first order S-matrix elements do not distinguish between $\mathscr{H}^{\text{p.v.}}$ and $\hat{\mathscr{H}}^{\text{p.v.}}$ with any r. Commutation relations are of no help either at least as long they are of the form of Eqs. (4) and (19).

ii) The S-Wave Amplitudes

For a discussion of the S-wave amplitudes we consider one particle matrix elements of the commutation relation (4), for example

$$\langle P|[F_5^{1+i2}, H^{\text{p.v.}}(t)]\,|\Lambda\rangle = \langle N|\,H^{\text{p.v.}}(t)|\Lambda\rangle\,.$$

On the left hand side the π-contribution (in disconnected parts [4]) and the neutron and Σ^+-intermediate states can be extracted:

$$\begin{aligned}
& i f_\pi \sqrt{2}(2\pi)^{3/2}\tfrac{1}{2}(\langle P\pi^-|\mathscr{H}^{\text{p.v.}}(0)|\Lambda\rangle + \langle P|\mathscr{H}^{\text{p.v.}}(0)|\Lambda\pi^+\rangle) \\
& + \langle P|A_0^{1+i2}(0)|N\rangle \frac{(2\pi)^3}{2p_N^0}\langle N|\mathscr{H}_{(0)}^{\text{p.v.}}|\Lambda\rangle \\
& - \langle P|\mathscr{H}_{(0)}^{\text{p.v.}}|\Sigma^+\rangle \frac{(2\pi)^3}{2p_{\Sigma^+}^0}\langle \Sigma^+|A_0^{1+i2}|\Lambda\rangle \\
& + \sum_Z \frac{\langle P|\int d^3x\,\partial^\lambda A_\lambda^{1+i2}(x)|Z\rangle\,\langle Z|\mathscr{H}^{\text{p.v.}}(0)|\Lambda\rangle}{i(p_p^0 - p_Z^0)} \\
& - \sum_{Z'} \frac{\langle P|\mathscr{H}^{\text{p.v.}}(0)|Z'\rangle\,\langle Z'|\int d^3x\,\partial^\lambda A_\lambda^{1+i2}(x)|\Lambda\rangle}{i(p_{Z'}^0 - p_\Lambda^0)} \\
& = -\langle N|\mathscr{H}^{\text{p.c.}}(0)|\Lambda\rangle\,.
\end{aligned} \tag{16}$$

In this equation the π-meson occurs with zero space momentum. The other particles and the intermediate states have the same space momenta as the Λ-particle. The S-wave decay amplitude appears in (16) at the two kinematical points

$$\begin{gathered}
s_{1,2} = M_p^2 + m_\pi^2 \pm 2p_p^0 m_\pi\,, \qquad u_{1,2} = M_\Lambda^2 + m_\pi^2 \mp 2p_\Lambda^0 m_\pi\,, \\
t_{1,2} = (p_\Lambda - p_p)^2\,.
\end{gathered} \tag{17}$$

The two points are closest to each other for $\boldsymbol{p}_\Lambda = \boldsymbol{p}_p = \boldsymbol{0}$ in agreement with the remarks made in the introduction. They approach each other only in the formal limit $m_\pi \to 0$. We choose therefore $\boldsymbol{p}_\Lambda = \mathbf{0}$ restricting thereby drastically the contributing intermediate states. Only $1/2^-$-states remain in the sum. In particular, the parts with the neutron and Σ^+ state in Eq. (16) vanish. The remaining contributions from $1/2^-$-states are proportional to the divergence of the axial vector current.

The right hand side in (16) is the "soft π-meson limit", for the kinematical configuration $\boldsymbol{p}_\Lambda = \boldsymbol{p}_\mathrm{p} = \mathbf{0}$. Indeed, the sum on the left vanishes in the limit of a strict conservation of the axial vector current and for $m_\pi = 0$. Thereby the transition matrix element and its π-meson crossed partner become identical. In the literature the soft limit has been discussed in the framework of generalized Ward-identities and by off-mass-shell considerations. Such treatments can be of formal help. It has been overlooked, however, that the *entire physical content of the formalism is contained in simple sum rules of the type of Eq.* (16) [8]. This sum rule gives the limit together with is corrections.

In our case the corrections are expected to be large. In fact, the terms on the left hand side of (16) are dominated by the K_1^0-pole. Important direct contributions from the K_1^0-particle arise through the disconnected $K_1^0 - P$ and $K_1^0 - \Lambda$ intermediate states etc. The right hand side of the commutation relation however, is presumably small since no scalar particle of low mass is known. Thus, it is necessary to take the K-dominated parts away. This can be achieved by using $\hat{\mathscr{H}}^{\mathrm{p.v.}}$ (Eq. 15) instead of $\mathscr{H}^{\mathrm{p.v.}}$. By this replacement the disconnected K-states drop out, the K-dominance is removed and the neglection of the remaining sum over intermediate 1/2-states compared to the right hand side of the commutation relation seems possible. The matrix elements of the current divergence $\partial^\lambda A_\lambda^{1+i2}(x)$ are expected to vanish rapidly with increasing momentum transfer (PCAC). Thus we obtain

$$\begin{aligned}
&\tfrac{1}{2}\{\langle p\pi^-|\hat{\mathscr{H}}^{\mathrm{p.v.}}|\Lambda\rangle + \langle P|\hat{\mathscr{H}}^{\mathrm{p.v.}}|\Lambda\pi^+\rangle\} \\
&\quad \cong -\frac{1}{i f_\pi \sqrt{2}(2\pi)^{3/2}}\{\langle N|\mathscr{H}^{\mathrm{p.c.}}|\Lambda\rangle - r\langle P|[F_5^{1+i2}, \partial^\lambda A_\lambda^7]|\Lambda\rangle\}, \qquad (18)\\
&\boldsymbol{p}_\Lambda = \boldsymbol{p}_\pi = \boldsymbol{p}_\mathrm{p} = \mathbf{0}\ .
\end{aligned}$$

The problem is now shifted to the commutator appearing on the right hand side. However, the quark model and other simple models in which the mass terms are responsible for $SU\,3$ and chiral symmetry breaking provides us with the necessary commutation relation:

$$\begin{aligned}
[F_5^\alpha(t), \partial^\lambda V_\lambda^7(0)]_{t=0} &= c\,[F^\alpha, \partial^\lambda A_\lambda(0)] \\
[F_5^\alpha(t), \partial^\lambda A_\lambda^7(0)]_{t=0} &= \frac{1}{c}\,[F^\alpha, \partial^\lambda V_\lambda^7(0)]\,, \qquad (19)\\
\alpha &= 1, 2, 3\,.
\end{aligned}$$

The constant c is expected to be smaller than 1.

Eq. (18) takes now the form

$$\frac{1}{2}\{\langle P\pi^-|\hat{\mathscr{H}}^{\text{p.v.}}|\Lambda\rangle+\langle P|\hat{\mathscr{H}}^{\text{p.v.}}|\Lambda\pi^+\rangle\} \simeq -\frac{1}{i f_\pi \sqrt{2}(2\pi)^{3/2}}\langle N|\hat{\mathscr{H}}^{\text{p.c.}}|\Lambda\rangle \tag{20}$$

with

$$\langle N|\hat{\mathscr{H}}^{\text{p.c.}}|\Lambda\rangle\equiv\langle N|\mathscr{H}^{\text{p.c.}}-\frac{r}{c}\,\partial^\lambda V_\lambda^7|\Lambda\rangle=\hat{a}(t)\left(-\frac{1}{\sqrt{3}}\hat{d}-\sqrt{3}\hat{f}\right)(\bar{u}_N u_\Lambda).$$

Replacing $\mathscr{H}^{\text{p.v.}}$ by $\hat{\mathscr{H}}^{\text{p.v.}}$ in Eq. (12) we obtain from (17), (18), (20) the following condition on the Λ°_--amplitudes:

$$\frac{1}{2}\hat{f}(s_1,t,u_1)+\frac{1}{2}\hat{f}(s_2,t,u_2)+m_\pi\frac{1}{2}(\hat{g}(s_1,t,u_1)-\hat{g}(s_2,t,u_2)) \simeq -\frac{1}{i f_\pi\sqrt{2}}\,\frac{1}{(2\pi)^{3/2}}\,\hat{a}(t)\left(\hat{d}\left(-\frac{1}{\sqrt{3}}\right)+\hat{f}(-\sqrt{3})\right). \tag{21}$$

The equation is valid for $t=(M_\Lambda-M_p)^2$.

The problem we are faced with now is how this result and the observable transition amplitude are to be related. We have to connect the points $s_{1,2}=(M_f\pm m_\pi)^2$, $t=(M_i-M_f)^2$, $u_{1,2}=(M_i\mp m_\pi)^2$ in s, t, u space with the decay point $s=M_i^2$, $t=m_\pi^2$, $u=M_f^2$. This can be done with the help of ordinary dispersion relations. However, we have to be careful and cannot put $m_\pi=0$ because baryon pole terms are present (for Λ-decay at $s=M_N^2$, $u=M_{\Sigma^+}^2$). The dependence on t is less critical since the distance between $t=(M_i-M_f)^2$ and $t=m_\pi^2$ is small and both points are far away from $t=m_{K^*}^2$. In the u, s-plane we have the picture:

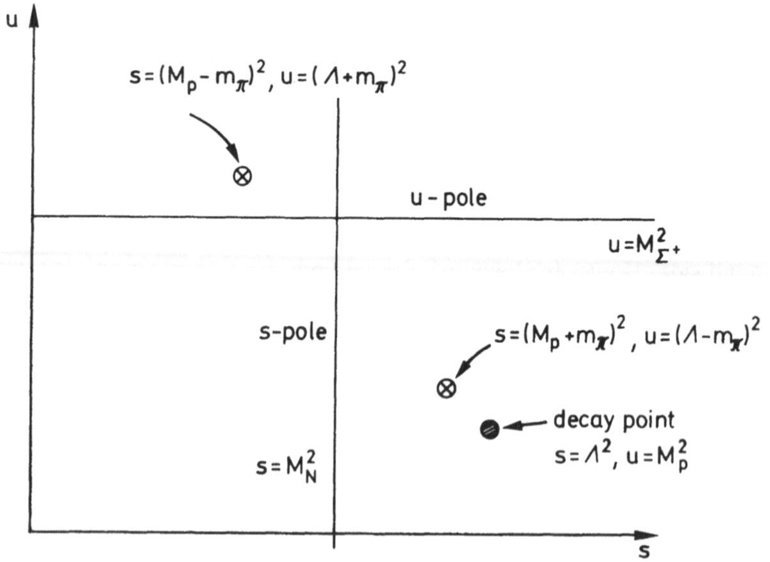

To describe the variation with s and u we take the graphs

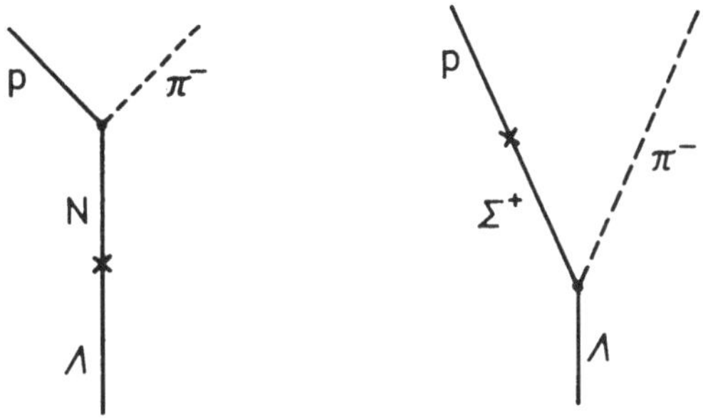

and similar graphs for the other decays. We write

$$\hat{f}(s,u) = f_0 + \hat{f}^{\text{pole}}(s,u) \tag{22}$$

and assume f_0 to be approximately constant in the region of interest. Eq. (22) corresponds to a subtracted dispersion relation. For $\hat{g}(s,u)$ we simply take the pole contribution

$$\hat{g}(s,u) = \hat{g}^{\text{pole}}(s,u)\,. \tag{23}$$

An unsubtracted dispersion relation is plausible from the definition of $g(s,t,u)$ in Eq. (12). The result from the commutator (Eq. (21)) can then be used to find f_0 in terms $\hat{a}(t)$. Finally we remember that at the decay point the relation

$$\hat{f} + (M_i - M_f)\,\hat{g} = f + (M_i - M_f)\,g = A$$

holds.

The pole contributions are easily calculated. For Λ°_{-}-decay, for example, one gets

$$\begin{aligned}
\hat{f}^{\text{pole}}(s,u) &= \frac{-g\hat{b}}{i(2\pi)^{3/2}}\,\frac{M_{\Sigma^+} - M_{\Lambda}}{M_{\Sigma^+}^2 - u}\,\frac{2\sqrt{2}}{\sqrt{3}}\,D(\hat{D} - \hat{F})\,, \\
\hat{g}^{\text{pole}}(s,u) &= \frac{g\cdot\hat{b}}{i(2\pi)^{3/2}}\left\{\frac{-\sqrt{2}(D+F)\left(\frac{1}{\sqrt{3}}\hat{D} + \sqrt{3}\hat{F}\right)}{M_{\mathrm{N}}^2 - s} - \frac{2\sqrt{\frac{2}{3}}\,D(\hat{D} - \hat{F})}{M_{\Sigma^+}^2 - u}\right\}.
\end{aligned} \tag{24}$$

The strong meson-baryon coupling constant has been denoted by $g(g^2/4\pi = 14.6)$. The D/F ratio of the one particle matrix element of $\mathscr{H}^{\text{p.v.}}$: $\hat{D}/\hat{F}$ is unknown. For the following estimate of the influence of the S-wave pole terms we use arbitrarily

$$\hat{F} = F = \tfrac{2}{5}, \qquad \hat{D} = D = \tfrac{3}{5}\,.$$

We are now able to express the S-wave amplitudes by the quantities $\hat{a}$, $\hat{d}/\hat{f}$ and $g\hat{b}$. We find with the help of the Eqs. (21)–(24) and similar formulas for the other decays:

$$\begin{aligned}
A(\Lambda^{\circ}_{-}) &= \frac{\hat{a}}{-i f_\pi \sqrt{2}(2\pi)^{3/2}}\left(-\frac{\hat{d}}{\sqrt{3}}-\sqrt{3}\,\hat{f}\right)+\frac{g\hat{b}}{i(2\pi)^{3/2}}\frac{1}{M_p}(-0.068)\,,\\
A(\Sigma^{+}_{\circ}) &= \frac{\hat{a}}{-i f_\pi \sqrt{2}(2\pi)^{3/2}}(-\hat{d}+\hat{f})+\frac{g\hat{b}}{i(2\pi)^{3/2}}\frac{1}{M_p}0.016\,,\\
A(\Sigma^{+}_{+}) &= 0+\frac{g\hat{b}}{i(2\pi)^{3/2}}\frac{1}{M_p}(-0.053)\,,\\
A(\Xi^{-}_{-}) &= \frac{a}{-i f_\pi \sqrt{2}(2\pi)^{3/2}}\left(-\frac{\hat{d}}{\sqrt{3}}+\sqrt{3}\,\hat{f}\right)+\frac{g\hat{b}}{i(2\pi)^{3/2}}\frac{1}{M_p}0.019\,.
\end{aligned} \tag{25}$$

The second term in each amplitude is the correction to our special soft π-meson limit due to the variation of the S-wave pole terms from the points $s_{1,2}$ and $u_{1,2}$ to the physical decay point. These corrections are small (if $\hat{b} \lesssim \hat{a}$) in spite of the large meson-nucleon coupling constant g. It is an essential point that $\hat{b}$ instead of b occurs. $\hat{b}$ is smaller than b and has no strong variation with momentum transfer in the region of interest. Unfortunately, we have no reliable estimate on its value. $\hat{b} \simeq a$ is suggested by $SU\,3 \times SU\,3$. Neglecting the correction terms we obtain:

$$A(\Sigma^{+}_{+}) = 0$$

and the Lee-Sugawara rule for S-waves

$$2A(\Xi^{-}_{-}) + A(\Lambda^{\circ}_{-}) = \sqrt{3}\,A(\Sigma^{+}_{\circ})$$

in very good agreement with experiment. We conclude from this agreement that one particle matrix elements of $\hat{\mathscr{H}}^{\text{p.v.}}$ show a strong octet dominance. The $\hat{d}/\hat{f}$ ratio is easily found from the experimental numbers and gives $\hat{d}/\hat{f} \simeq -1/3$ [9].

This result, octet dominance and $d/f \simeq -1/3$ is indeed what is predicted from the *second* term in the expression

$$\hat{\mathscr{H}}^{\text{p.c.}} = \mathscr{H}^{\text{p.c.}} - \frac{r}{C}\,\partial^{\lambda} V^{7}_{\lambda}$$

as can be seen by using $SU\,3$ F-type coupling for the matrix elements of the charge density V^{7}_{0} [10]. The first term ($\mathscr{H}^{\text{p.c.}}$) would have the same behaviour in case a spin 0 particle (κ-meson) dominates the dispersion relation for the matrix element of both operators. However, it seems somewhat more natural to assume that the second term is the significant one. If we accept this point of view the measured values of the S-wave decay amplitudes allow a determination of r/c and give only an upper limit for the one particle matrix elements of $\mathscr{H}^{\text{p.c.}}$. This should be kept in mind in the treatment of p-wave decays where these matrix elements enter. At present the p-wave decay amplitudes are not properly understood and we will not discuss them.

3. Octet Dominance and the Structure of Weak Interaction

We have seen that the one-particle matrix elements of the effective parity conserving weak Hamiltonian $\hat{\mathscr{H}}^{\text{p.c.}}$ show octet dominance. It was particularly remarkable that the $\hat{d}/\hat{f}$ ratio found agreed in magnitude and sign with the d/f ratio of the $SU\,3$ mass-splitting Hamiltonian. We may also take a look at the electromagnetic mass differences. Here octet dominance is also observed but not as pronounced as in strong mass-splitting and weak interactions. Nevertheless, the enhanced octet has a ratio $d/f \approx -1/2$ in qualitative agreement with the value found in the other two interactions [10]. *We conclude that the one-particle matrix elements of the dominant octet contributions in parity conserving weak* $\hat{\mathscr{H}}^{\text{p.c.}}$, *electromagnetic and strong interactions are apart from* $SU\,3$ *rotations approximately proportional to each other.* The proportionality gives rise to a large number of formulas connecting ratios of mass differences with each other and with ratios of S-Wave decay amplitudes. A further test of the proportionality is obtained by comparing the ratios between meson- and corresponding baryon-matrix elements of the three interactions [9]. A typical relation obtained in this way is (we take $d/f = -1/3$).

$$\frac{M_K^2 - M_\pi^2}{M_\Sigma^2 - M_N^2} \approx \frac{M_{K^0}^2 - M_{K^+}^2}{2(M_N^2 - M_P^2)} \approx \frac{A(K_s^0 \to \pi^+ + \pi^-)}{A(\Lambda \to P + \pi^-)_s}\,\frac{2}{\sqrt{3}}, \tag{26}$$

$$0.42 \approx 0.82 \approx 0.65\,,$$

where the numbers in the second line indicate the experimental values. The K-meson mass difference is predicted in terms of the neutron-proton mass difference to about 50%. This value is improved if we subtract from the electromagnetic mass differences the nonenhanced parts. The relation also predicts the $K_s^0 \to \pi^+ + \pi^-$ decay amplitude within 35% in terms of the S-wave Λ-decay amplitude. In view of the fact that we compare one-particle matrix elements and consequently had to use soft pion formulas also for K-decay this result is satisfactory. We like to emphazise that formulas like Eq. (26) are according to our discussion simply due to the proportionality of the octet parts of one-particle matrix elements and thereby independent of special model assumptions as sometimes proposed. In particular the sign and rough magnitude of all electromagnetic $\Delta I = 1$ mass difference is known as soon as one mass difference is taken from experiment.

The common feature of electromagnetic, effective parity conserving weak ($\hat{\mathscr{H}}^{\text{p.c.}}$) and strong interactions may be due to the *singular nature of the product of two currents at small space-time distances.* This idea is supported by the observation that octet dominance is less pronounced in electromagnetic interactions where short range contributions are less important. Octet dominance is more striking in the short range weak interaction and strong interaction (which is perhaps also of a current × current form). We may also mention that our inability to compute correctly the octet part of electromagnetic mass differences from low lying states indicates the presence of short range effects which are more difficult to handle.

If small space-time distances of the current product are indeed responsible for octet dominance the structure of weak interactions could become important for the description of nonleptonic decays. We know nothing about this structure except that it is of short range. We will now compare the short range contribution of weak and electromagnetic matrix elements. To do this we have to make a number of drastic assumptions. This last part of our lecture is therefore purely speculative.

The second order electromagnetic interaction Hamiltonian is from perturbation theory

$$\mathscr{H}_{\text{el}}(0) = \frac{e^2/4\pi}{(2\pi)^3} \int \frac{d^4 q}{-q^2 - i\varepsilon} g^{\lambda\lambda'} \times \int d^4 x \, e^{iqx} \left[\tau\left(J_\lambda^{\text{el}}\left(\frac{x}{2}\right) J_{\lambda'}^{\text{el}}\left(-\frac{x}{2}\right)\right) + \text{Schwinger terms} \right]. \tag{27}$$

The symbol τ stands for time ordering.

The Schwinger terms are model dependent gauge contributions supplementing the time-ordered product of the current such that a covariant and gauge invariant expression results. We will assume that these extra terms do not contribute to the differences of the masses.

The neutron-proton mass difference, for example, may then be written in the form

$$\int \frac{d^4 q}{-q^2 - i\varepsilon} g^{\lambda\lambda'} T_{\lambda\lambda'}^{\text{el}}(q, p) = \frac{M_{\text{N}} - M_{\text{P}}}{e^2/4\pi} M_{\text{P}} \sqrt{3} \tag{28}$$

where

$$T_{\lambda\lambda'}^{\text{el}}(q, p) = \int d^4 x \, e^{iqx} \langle N | \tau \frac{1}{2} \left\{ V_\lambda^{1-i2}\left(\frac{x}{2}\right), V_{\lambda'}^{8}\left(-\frac{x}{2}\right) \right\} | P \rangle . \tag{29}$$

A typical weak interaction matrix element is

$$\frac{1}{M_p^2}\int d^4q S^{\lambda\lambda'}(q)\, T^{\text{weak}}_{\lambda\lambda'}(q,p) = \frac{(2\pi)^4 \langle N|\mathscr{H}^{\text{p.c.}}_{\text{weak}}|\Lambda + \frac{1}{\sqrt{3}}\Sigma^\circ\rangle}{\sin\theta\cdot\cos\theta\cdot \mathrm{G}\cdot M_p^2/\sqrt{2}} \lesssim 0.21\, M_p^2 \tag{30}$$

with

$$T^{\text{weak}}_{\lambda\lambda'}(q,p) = \int d^4x\, e^{iqx}\langle N|\tau\left[\frac{1}{2}\left\{V_\lambda^{1-i2}\left(\frac{x}{2}\right), V_{\lambda'}^{4+i5}\left(-\frac{x}{2}\right)\right\}_+ + \frac{1}{2}\left\{A_\lambda^{1-i2}\left(\frac{x}{2}\right), A_{\lambda'}^{4+i5}\left(-\frac{x}{2}\right)\right\}_+\right]|\Lambda + \frac{1}{\sqrt{3}}\Sigma^\circ\rangle. \tag{31}$$

The numerical value has been obtained from the analysis of S-wave decay amplitudes performed in Chapter 1. We used the value for $\langle\hat{\mathscr{H}}^{\text{p.c.}}\rangle$ which is presumably an upper limit for $\langle\mathscr{H}^{\text{p.c.}}\rangle$. In Eq. (30) we introduced a structure function $S^{\lambda\lambda'}(q)$ with the property

$$S^{\lambda\lambda'}(0) = g^{\lambda\lambda'}.$$

Now we make the assumption that the integral over q in Eq. (30) has the *same convergence property* as the q-integral in the electromagnetic case. According to *Björkén* [11] the latter integral is likely to be logarithmically divergent. A cut-off has to be introduced if this is true.

Two essentially different possibilities may now be discussed:

i) $T^{\text{weak}}_{\lambda\lambda'}$ has the same asymptotic behaviour as a function of q as $T^{\text{el}}_{\lambda\lambda'}$.

To insure equal convergence properties the structure function $S^{\lambda\lambda'}(q)$ has to decrease $\sim 1/q^2$ in this case. We take

$$S^{\lambda\lambda'}(q) = \frac{M_B^2}{M_B^2 - q^2}\, g^{\lambda\lambda'} \tag{32}$$

and regard this form as the simplest one-parameter description of a presumably more complicated structure of the weak interaction. An intermediate boson with spin 1 would give an additional term $-q^\lambda q^{\lambda'}/M_B^2$ added to $g^{\lambda\lambda'}$. We disregard this term because of our ignorance about possible compensating gauge type parts, but mainly because its asymptotic behaviour is not in agreement with our assumption. Next we suppose that the octet parts in both interactions are obtained from high q values of the integrals as suggested by our previous discussion. For such large q-values we may define a quantity Z by

$$Z = \frac{-g^{\lambda\lambda'}\, T^{\text{weak}}_{\lambda\lambda'}}{2\sqrt{2}\, g^{\lambda\lambda'}\, T^{\text{el}}_{\lambda\lambda'}}. \tag{33}$$

Eq. (28) and (30) give

$$\frac{M_B^2}{M_P^2}|Z| \lesssim 0.15\,. \tag{34}$$

We took for the short range octet part of the neutron-proton mass difference $(M_N - M_P)_8 \simeq 2$ MeV. The low number on the right hand side of Eq. (34) is remarkable. In strict $SU\,3$ and for $\langle VV\rangle = \langle AA\rangle$ the quantity Z would be 1. Symmetry breaking effects may reduce it to 1/10 or so, but certainly not to a much smaller value. We conclude that the *parameter M_B representing the structure of weak interaction has a surprisingly low value $M_B \lesssim M_P$*. Since $M_B < M_P$ can not be accepted a smaller value for $\langle \mathscr{H}^{\text{p.c.}}\rangle$ as the one used could easily rule out case i).

Let us therefore discuss a second possibility by assuming:

ii) $T_{\lambda\lambda'}^{\text{weak}}$ decreases faster than $T_{\lambda\lambda'}^{\text{el}}$ as a function of the variable q. According to the method of *Björkén* we then expect an additional factor $1/q^2$. One may write in this case

$$T_{\lambda\lambda'}^{\text{weak}} \Rightarrow 2\sqrt{2}\,\tilde{Z}\,\frac{M_P^2}{q^2}\,T_{\lambda\lambda'}^{\text{el}}\,. \tag{35}$$

The asymptotic validity of $SU\,3 \times SU\,3$ may be the reason for this fast decrease. Now the structure function $S^{\lambda\lambda'}(q)$ is unimportant and may be replaced by the point interaction form $g^{\lambda\lambda'}$. The numerical comparison of Eq. (28) and (30) gives again the small number $|\tilde{Z}| \lesssim 0.15$ indicating an interesting suppression of the weak matrix element.

The main part of this work was done during the author's stay at the Nordisk Institute for Theoretical Atomphysics (Nordita) in Copenhagen. He would like to express his graditude to Prof. *J. Hamilton* for his kind hospitality and for useful discussions. He also profited from discussions with Dr. *K. Hefft* and *S. Minutschehr*.

References and Remarks

1. General References: *Bludmann, S. A.*: Cargése Lectures in Physics. New York: Gordon and Breach 1967. — *Struminski, B. V.*: Ecole Internationale Herceg Novi (Yougoslavie) (1967).
2. *Stech, B.*: Lectures on weak decays of hadrons Nordita, Copenhagen, 1968. This paper describes in addition the general features of semileptonic decays. The part on nonleptonic decays and mass-differences is practical identical to this lecture.
3. In a commutator containing a charge operator and a non-local interaction the time dependence of the charge operators has to be considered and compared with the effective time differences in the interaction. Corresponding modifications are most important for the long range electromagnetic interaction. We believe that the troubles one has in the description of η-decay are due to this effect.

4. *Fubini, S.*, and *G. Furlan:* Dispersion theory of low-energy limits. Ann. Phys. (N.Y.) **48**, 2 (June 1968).
— Nobel Symposium, Aspenäsgården (1968) to be published.
5. This concept has successfully been used for the case case of the K_{l_3} formfactors by *Dahmen, H., K. Rothe*, and *L. Schülke:* Heidelberg preprint 1968.
6. *Lobashov, V. M., V. A. Nazarenko, L. F. Saenko, L. M. Smotritskii*, and *G. I. Kharkevich:* Zh. Eksper. Teor. Fiz. **5**, 73 (1967); Engl. Trans. JETP Letters **5**, 59 (1967). — *Avov, Yu. G., P. A. Kruptchitsky, M. Y. Bulgakov, O. N. Yermakov*, and *I. L. Karpikhin:* Phys. Letters **27** B, 16 (1968). Earlier references are given in these articles.
7. We use invariant state normalization

$$\langle p'|p\rangle = 2p^0\,\delta^3(p'-p),$$
$$\langle|S-1|\rangle = i\,\delta^4(p_f-p_i)\,T,$$
$$\bar{u}u = 2M.$$

8. I call this the "no wonder" theorem.
9. *Müller, V. F.*, and *B. Stech:* Nuclear Phys. B **3**, 464 (1967).
10. *Hefft, K.*, and *B. Stech:* Z. Physik **202**, 514 (1967).
11. *Björkén, J. D.:* Phys. Rev. **148**, 1467 (1966).

Prof. Dr. *Berthold Stech*
Institut für Theoretische Physik
der Universität Heidelberg
6900 Heidelberg, Philosophenweg 16

Current Algebra in the Framework of General Quantum Field Theory

P. STICHEL

Contents

1. Introduction

The incorporation of current algebra as a dynamical principle into the general framework of quantum field theory gives rise to the following questions:

1. What is the correct formulation of equal-time commutation relations (ETCR) in field theory, in particular

a) how can one define charges respectively generalized charges;

b) what is the precise meaning of "equal times".

2. Are there any restrictions on the form of ETCR which follow already from the general axioms of quantum field theory.

Concerning this last point not much is known up to day. We have only some general results on the vacuum-expectation values (VEV) of equal-time commutators (ETC) – this is related to the problem of gradient terms in ETCR.

Before going into the details I just want to remind you on the main axioms of field theory [1]:

1. Relativistic covariance.
2. Spectral condition.

3. Field operators $A_i(x)$ are operator valued tempered distributions. The smeared fields $A_i(f)$ are defined on a dense set of states in Hilbert-space $\mathfrak{H}$.
4. Locality.
5. Asymptotic completeness ($\mathfrak{H} = \mathfrak{H}^{\text{in}} = \mathfrak{H}^{\text{out}}$).

2. Definition of Charges and Generalized Charges

Charges respectively generalized charges Q are formally defined as space integrated time components of a conserved respectively non-conserved current

$$\text{“}Q\text{”} = \int d^3x j_0(x). \tag{1}$$

Eq. (1) makes no sense mathematically because current operators arc operator-valued distributions. Therefore, only smeared currents $j_\mu(f)$

$$j_\mu(f) = \int d^4x f(x) j_\mu(x)$$

with* $f \in \mathscr{S}$ exist as unbounded operators.

In particular let us consider for $f(x)$ a product of a space- and a time-smearing function

$$f(x) = f_R(|\boldsymbol{x}|)\, f_T(x_0)$$

with

$$\{f_R(|\boldsymbol{x}|)\} \xrightarrow[R\to\infty]{} 1.$$

Then our problem of the definition of a charge respectively generalized charge may be stated as follows: In what sense exists the limit of the sequence of unbounded operators

$$\lim_{R\to\infty} j_0(f_R, f_T). \tag{2}$$

We suppose that our theory contains no zero-mass particles. Then the answer to our problem looks as follows [2]:

a) If j_μ is a conserved current, then $\{j_0(f_R, f_T)\} \xrightarrow[R\to\infty]{} Q$ on the dense set of quasilocal states, where Q is independent of the sequence $\{f_R\}$. Q is called the charge operator.

b) If j_μ is a non-conserved current, then the sequence $\{j_0(f_R, f_T)\}$ converges for $R\to\infty$ weakly on the dense set of quasi-local states without defining an operator, i.e. a generalized charge only defines a bilinear form.

* $\mathscr{S}$ is the set of functions which are infinitely often differentiable and decrease for $|x|\to\infty$ faster than any inverse polynomial.

Before discussing the proofs of the statements made above, we have to explain some notions:

Strong Convergence

$$\{A_n\} \underset{n\to\infty}{\Rightarrow} A : \lim_{n\to\infty} \|(A_n - A)\,|\psi\rangle\| = 0$$

for every ψ out of a dense set in $\mathfrak{H}$.

Weak Convergence

$$\{A_n\} \xrightarrow[n\to\infty]{} A : \lim_{n\to\infty} \langle\psi|(A_n - A)|\phi\rangle = 0$$

for every ψ and ϕ out of a dense set in $\mathfrak{H}$.

Quasi-Local States

A state $|\phi\rangle$ is called quasi-local if it has the form

$$|\phi\rangle = \sum_{m=1}^{N} \int d^4x_1 \ldots d^4x_m g_m(x_1 \ldots x_m) \cdot A_1(x_1) \ldots A_m(x_m), \tag{3}$$

where the A's are from the basic set of local fields and $g_m \in \mathscr{S}$. If we add to the set of all quasi-local states the vacuum, we then have a dense set of states $\mathfrak{H}' \subset \mathfrak{H}$.

We will now prove a set of statements which is equivalent to a) and b).

Statement 1

$$\{j_0(f_R, f_T)\} \not\Rightarrow Q .$$

Proof. Suppose we have $\{j_0(f_R, f_T)\} \underset{R\to\infty}{\Rightarrow} Q$ on a dense set of states including the vacuum. Then we must have at least $\|(j_0(f_R, f_T) - Q)\,|0\rangle\| \xrightarrow[R\to\infty]{} 0$. Necessary for that is the existence of

$$\lim_{R\to\infty} \langle 0|\, j_0(f_R, f_T)\, j_0(f_R, f_T)\,|0\rangle . \tag{4}$$

By using for the VEV of the product of two current operators the Källén-Lehmann-representation (compare Chapter 4) it is a matter of straightforward computation to obtain for the limit in (4)

$$\begin{aligned} &cR^2 \quad \text{for} \quad \partial^\mu j_\mu = 0 \\ &c'R^3 \quad \text{for} \quad \partial^\mu j_\mu \neq 0 \end{aligned}$$

in case of the particular sequence

$$f_R(|\boldsymbol{x}|) = \begin{cases} 1 & |\boldsymbol{x}| < R \\ 0 & |\boldsymbol{x}| > R + L . \end{cases} \tag{5}$$

Therefore, we can expect at most $\{j_0(f_R, f_T)\} \xrightarrow[R\to\infty]{} Q$.

Statement 2

The bilinear form

$$L(\psi, \phi) = \lim_{R\to\infty} \langle\psi|j_0(f_R, f_T)|\phi\rangle$$

exists for every $\psi, \phi \in \mathfrak{H}'$ and is independent of the sequence $\{f_R\}$.

Proof. The statement follows immediately from the result of *Ruelle* [3], that $\langle\psi|j_\mu(\boldsymbol{x}, t)|\phi\rangle$ is a smooth function in $\boldsymbol{x}$ which goes for $|\boldsymbol{x}|\to\infty$ to zero faster than any inverse polynomial of $|\boldsymbol{x}|$ for every $\psi, \phi \in \mathfrak{H}'$.

Because we cannot proof the existence of $L(\psi, \phi)$ for arbitrary normalizable states (there exist counter-examples) we can expect $j_0(f_R, f_T) \xrightarrow[R\to\infty]{} Q$ at most on $\mathfrak{H}'$.

From the existence of $L(\psi, \phi)$ we conclude that $\{j_0(f_R, f_T)\} \xrightarrow[R\to\infty]{} Q$ if and only if $L(\psi, \phi)$ is bounded with respect to ψ respectively ϕ:

$$\left.\begin{array}{ll} & |L(\psi, \phi)| \leqq \|\phi\| \; c(\psi) \\ \text{respectively} & |L(\psi, \phi)| \leqq \|\psi\| \; c'(\phi) \end{array}\right\} \text{ for every } \quad \psi, \phi \in \mathfrak{H}' .$$

Statement 3

$$\{j_0(f_R, f_T)\} \xrightarrow[R\to\infty]{} Q \quad \text{on} \quad \mathfrak{H}' \quad \text{if} \quad \partial^\mu j_\mu = 0 .$$

Proof. 1. If Statement 3 is correct, we must have due to translational invariance and $\langle 0|j_\mu(0)|0\rangle = 0$ that $\lim_{R\to\infty} \langle\phi|j_0(f_R, f_T)|0\rangle = 0$ for every quasi-local state ϕ. We now use the fact that due to the absence of zero-mass particles, the state $|\psi\rangle = \frac{1}{H}|\phi\rangle$ is again a quasi-local one. Therefore, by means of the continuity equation for j_μ we have

$$\langle\phi|j_0(f_R, f_T)|0\rangle = \langle\psi|[H, j_0(f_R, f_T)]|0\rangle = -i\langle\psi|j_r(f_R', f_T)|0\rangle , \quad (6)$$

where j_r is the component of j_μ along the radius vector $\boldsymbol{x}$. Because $\{f_R'\} \xrightarrow[R\to\infty]{} 0$ the r.h.s. of (6) converges due to statement 2 for $R\to\infty$ to zero.

2. Now we will show that

$$\lim_{R\to\infty} \langle\psi|j_0(f_R, f_T)|\phi\rangle \quad (7)$$

is bounded e.g. with respect to ψ. We write $|\phi\rangle = B|0\rangle$ where B is a quasi-local operator. Then (7) may be written as follows

$$\lim_{R\to\infty} \langle\psi|[j_0(f_R, f_T), B]|0\rangle + \lim_{R\to\infty} \langle\psi|Bj_0(f_R, f_T)|0\rangle . \quad (8)$$

The second term in (8) vanishes according to argument 1 whereas the first term has the desired property, because $\lim_{R\to\infty} [j_0(f_R, f_T), B]$ is again a quasi-local operator.

Statement 4

$$\{j_0(f_R, f_T)\} \underset{R\to\infty}{\not\longrightarrow} Q \quad \text{on} \quad \mathfrak{H}' \quad \text{if} \quad \partial^\mu j_\mu = A \neq 0\,.$$

Proof. It is sufficient to give just a counter example: Consider the sequence of quasi-local states

$$|\phi_\varrho\rangle = \|A(f_\varrho, f_T)|0\rangle\|^{-1}\, H A(f_\varrho, f_T)|0\rangle$$

where we choose for f_ϱ the sequence (5).

By means of the *Källén-Lehmann* representation [6] one easily finds

$$\lim_{\varrho\to\infty}\,\lim_{R\to\infty} \frac{\langle\phi_\varrho|j_0(f_R, f_T)|0\rangle}{\|\phi_\varrho\|} = O(\varrho^{3/2}),$$

i.e. we have no boundedness with respect to ϕ_ϱ.

3. Formulation of Equal-Time Limits

3.1 ETC for Charges Respectively Generalized Charges

Statement. The truncated commutator of time smeared charges exists as a bilinear form $L(\psi, \phi)_g$ in $\mathfrak{H}'$

$$L(\psi, \phi)_g = \int d^3x \int d^3y \int dx_0 \int dy_0\, g(x_0, y_0)\langle\psi|\,[j_{0,i}(x), j_{0,k}(y)]\,|\phi\rangle_{\text{tr.}}$$

with $g \in \mathscr{S}$.

Proof. According to a theorem by *Ruelle* [3]

$$\int dx_0 \int dy_0\, g(x_0, y_0)\langle\psi|j_{0,i}(x)\, j_{0,k}(y)|\phi\rangle_{\text{tr.}}$$

is a smooth function of $\boldsymbol{x}$ and $\boldsymbol{y}$ which decreases to zero faster than any inverse polynomial if $|\boldsymbol{x}|$ respectively $|\boldsymbol{y}| \to \infty$ for every $\psi, \phi \in \mathfrak{H}'$. Therefore, we may integrate over the whole space in $\boldsymbol{x}$ respectively $\boldsymbol{y}$ without introducing first sequences of test functions $\{f_R\}$.

By specializing $g(x_0, y_0) = \varphi\left(\frac{x_0 + y_0}{2}\right) f_T(x_0 - y_0)$ and choosing for $\{f_T\}$ a sequence of test functions which for $T\to 0$ approaches the δ-function, the requirement for the existence of an equal-time limit of $L(\psi, \phi)_g$ may be formulated as follows:

The limit

$$\lim_{T\to\infty} L(\psi, \phi)_{\varphi\cdot f_T}$$

exists for all quasilocal states ψ, ϕ and is independent of the δ-sequence $\{f_T\}$ out of a class of admissible δ-sequences which will be specified in Section 3.3.

3.2 ETC for Currents

The requirement for the existence of an equal-time limit for current commutators may be formulated in analogy to the case of charges as follows: The quantity

$$\lim_{T\to 0}\int d^4x\int d^4y\,\varphi\left(\frac{x_0+y_0}{2}\right) f_T(x_0-y_0)\, f(\mathbf{x},\mathbf{y}),$$

$$\langle\psi|\,[j_{\mu,\alpha}(x), j_{\nu,\beta}(y)]\,|\phi\rangle \quad \text{with} \quad f\in\mathscr{S}$$

exists for all quasi-local states ψ, ϕ and is independent of the δ-sequence $\{f_T\}$ out of a class of admissible δ-sequences which will be specified in Section 3.3.

3.3 Choice of Admissible δ-Sequences $\{f_T\}$

We have still to specify the class of δ-sequences $\{f_T\}$ to be used for the definition of equal-time limits.

At first one might think that every sequence $\{f_T\}$ with $f_T\in\mathscr{S}$ converging in the topology of $\mathscr{S}'$ against a δ-function is admissible. But because we required the independence of the equal-time limit of the considered distribution $F(x_0-y_0)$ on the chosen sequence $\{f_T\}$, F must be itself a testfunction with $F\in\mathscr{S}$ [4]. This is not the case even in low order perturbation theory.

A natural choice for admissible sequences $\{f_T\}$ which seems to be appropriate for the known applications of current algebra (derivation of sum rules resp. low-energy theorems) has been proposed by *O. Steinmann* [4]: In momentum space we demand

$$\frac{\tilde{f}_T(q_0)}{(1+|q_0|)^{\varepsilon}} \xrightarrow[T\to 0]{} \frac{1}{(1+|q_0|)^{\varepsilon}}$$

and for the n-th derivative

$$\tilde{f}_T^{(n)}(q_0)\,(1+|q_0|)^{n-\varepsilon} \xrightarrow[T\to 0]{} 0 \quad \text{with} \quad n=1,2,\ldots$$

for every $\varepsilon>0$, where the convergence should be uniform in q_0. It turns out that test functions $\tilde{f}\in\mathscr{S}$ are dense in the function space spanned by all test functions $\tilde{f}$ which satisfy such a convergence criterion [4].

A particular class of sequences which converges in this sense to 1 is given by

$$\tilde{f}_T(q_0) = f(Tq_0)$$

with $f(0)=1$ and $f \in \mathscr{S}$ [4].

The class of δ-sequences defined in this way is larger than the one discussed in a recent paper [5]. Therefore, the requirement of the existence of an equal-time limit becomes more restrictive.

This rigorous definition of ETC will be applied in the next section to the case of the VEV of current commutators and in the next lecture to the investigation of ETCR in perturbation theory.

4. Gradient Terms

4.1 c-Number Gradient Terms

In order to investigate c-number gradient terms in ETCR of currents we have just to consider VEV of current commutators

$$\langle 0|\,[j_{\mu,i}(x), j_{\nu,k}(y)]\,|0\rangle \tag{1}$$

where both currents are of the same kind (vector resp. axialvector currents).

As a consequence of Lorentz-covariance and spectrum condition expression (1) satisfies the *Källén-Lehmann* spectral representation [6]:

$$-i(2\pi)^3 \int d\varkappa^2 \left(\varrho_1^{ik}(\varkappa^2)\, g_{\mu\nu} + \frac{\varrho_2^{ik}(\varkappa^2)}{\varkappa^2}\, \partial_\mu \partial_\nu \right) \Delta_\varkappa(x-y) \tag{2}$$

where the spectral functions ϱ_r^{ik} are defined by

$$\begin{aligned} \varrho_{\mu\nu}^{ik}(p) &= \sum_n \langle 0|j_{\mu,i}(0)\,|n\rangle \langle n|j_{\nu,k}(0)\,|0\rangle\, \delta(p^2 - p_n^2) \\ &= -g_{\mu\nu}\varrho_1^{ik}(p^2) + \frac{p_\mu p_\nu}{p^2}\, \varrho_2^{ik}(p^2)\,. \end{aligned} \tag{3}$$

Due to the definition of the commutator and (2) the ϱ_r^{ik} are symmetric

$$\varrho_r^{ik} = \varrho_r^{ki}\,.$$

In particular if i resp. k are isospin indices, we have

$$\varrho_r^{ik} = \delta_{ik}\varrho_r$$

as a consequence of isospin conservation.

If at least one of the two currents is conserved we obtain $p^\mu \varrho_{\mu\nu}(p) = 0$ and, therefore,

$$\varrho_1 = \varrho_2\,.$$

In the general case we have

$$\varrho_2^{ii} \geqq \varrho_1^{ii} \geqq 0 . \tag{4}$$

Proof of Inequality (4)

We suppose the current operators to be hermitean.

a) Consider Eq. (3) for $\mu = \nu = 3$ in a Lorentz-frame where $p_3 = 0$. Then we obtain

$$\varrho_1^{ii}(p^2) = \varrho_{33}^{ii}(p) \geqq 0 .$$

b) From Eq. (3) we infer

$$0 \leqq p^\mu p^\nu \varrho_{\mu\nu}^{ii}(p) = p^2(\varrho_2^{ii}(p^2) - \varrho_1^{ii}(p^2)) . \tag{5}$$

Due to $p^2 > 0$ (spectrum condition) (5) is equivalent to

$$\varrho_2^{ii}(p^2) - \varrho_1^{ii}(p^2) \geqq 0 .$$

Equal-Time Commutators

Model independent conclusions on ETC are based on the use of inequality (4), i.e. nothing can be said if $i \neq k$. Therefore, we consider exclusively the case $i = k$ (hereafter the index $i = k$ will be suppressed).

According to the definition of ETC's given in Chapter 3 we have to consider

$$\lim_{T \to 0} \int d^4x \, \varphi(x) \, f_T(x_0) \langle 0 | [j_\mu(x/2), j_\nu(-x/2)] | 0 \rangle$$

respectively according to (2)

$$-i(2\pi)^3 \lim_{T \to 0} \int d\varkappa^2 \int d^4x \, \varphi(x) \, f_T(x_0) \, [\varrho_1(\varkappa^2) \, g_{\mu\nu} + \frac{\varrho_2(\varkappa^2)}{\varkappa^2} \partial_\mu \partial_\nu] \, \Delta_\varkappa(x) . \tag{6}$$

a) $\mu = 0, k \neq 0$.

The ETC exists if and only if $\int d\varkappa^2 \frac{\varrho_2(\varkappa^2)}{\varkappa^2} < \infty$ and is given by

$$i(2\pi)^3 \int d^3x \, \varphi(\boldsymbol{x}) \, \partial_k \, \delta(\boldsymbol{x}) \int d\varkappa^2 \frac{\varrho_2(\varkappa^2)}{\varkappa^2} . \tag{7}$$

Proof. At first we rewrite (6) for $\mu = 0$, $\nu \neq 0$ into a momentum space integral

$$\lim_{T \to 0} \int d\varkappa^2 \frac{\varrho_2(\varkappa^2)}{\varkappa^2} \int d^3q \, q_\nu \, \tilde{f}_T^e(\sqrt{\boldsymbol{q}^2 + \varkappa^2}) \, \tilde{\varphi}(\boldsymbol{q}) , \tag{8}$$

where $\tilde{f}_T^e$ is the even part of $\tilde{f}_T$.

a) In order to show the non-existence of the ETC in the case of $\int d\varkappa^2 \frac{\varrho_2(\varkappa^2)}{\varkappa^2} = \infty$ it is according to Section 3.3 sufficient to prove the divergence of the ETC for one particular admissible δ-sequence $\{f_T\}$. We choose

$$\tilde{f}_T(q_0) = e^{-T^2 q_0^2}.$$

For this choice of f_T the existence of (8) is equivalent to the existence of

$$\lim_{T\to 0} \int d\varkappa^2 \frac{\varrho_2(\varkappa^2)}{\varkappa^2} e^{-T^2\varkappa^2}. \tag{9}$$

Because $\frac{\varrho_2(\varkappa^2)}{\varkappa^2} \geqq 0$ and $e^{-T^2\varkappa^2}$ is a positive and monotonic decreasing function of T we may interchange the limit $T\to 0$ and the $\varkappa^2$-integration, i.e. (9) diverges if $\int d\varkappa^2 \frac{\varrho_2(\varkappa^2)}{\varkappa^2}$ diverges.

b) Now we consider the case $\int d\varkappa^2 \frac{\varrho_2(\varkappa^2)}{\varkappa^2} < \infty$. Because $\tilde{f}_T^e(q_0)$ is bounded

$$|\tilde{f}_T^e(q_0)| \leqq 1,$$

we are allowed, due to Lebesque's bounded convergence criterion, to interchange $\lim T\to 0$ with the integrations in (8). In this way we arrive at (7). From (7) and the inequalities (4) we conclude that the ETC vanishes if and only if

$$\varrho_2(\varkappa^2) = \varrho_1(\varkappa^2) = 0.$$

By means of the Källén-Lehmann representation we then obtain

$$\langle 0| j_{\mu,i}(f)\, j_{\mu,i}(f)|0\rangle = 0, \qquad \forall f \in \mathscr{S}. \tag{10}$$

Due to the positivity of the norm in Hilbert-space (10) is equivalent to

$$j_{\mu,i}(x)|0\rangle = 0 \tag{11}$$

which implies according to the *Schroer-Jost-Federbush-Johnson* theorem [7]

$$j_{\mu,i}(x) = 0$$

if our currents are supposed to be local operators.

In this way we arrive at the final conclusion that the ETC of a time and a space component of one current contains, if it exists, a non-vanishing c-number gradient term [8].

b) $\mu = \nu = 0$.

The ETC is either zero or it does not exist.

Proof. In this case (6) looks in momentum space as follows

$$-\lim_{T\to 0} \int d\varkappa^2 \int d^3q\, \tilde{\varphi}(\boldsymbol{q}) \left(\varrho_1(\varkappa^2) - \frac{\varrho_2(\varkappa^2)(\boldsymbol{q}^2+\varkappa^2)}{\varkappa^2} \right) \cdot \frac{\tilde{f}_T^a(\sqrt{\boldsymbol{q}^2+\varkappa^2})}{\sqrt{q^2+\varkappa^2}}, \tag{12}$$

where $\tilde{f}_T^a$ is the antisymmetric part of $\tilde{f}_T(\{f_T^a\} \to 0)$. (12) may turn out to be different from zero if the $\varrho_r(\varkappa^2)$ have a bad asymptotic behaviour. Such a case means that the ETC does not exist, as every symmetric f_T leads to the vanishing of (12) already for a finite T.

c) μ, ν spacelike.

The whole discussion for the case $\mu = \nu = 0$ may be repeated.

4.2 *q*-Number Gradient Terms

Up to day it has not been possible to derive any model independent result on q-number gradient terms. Model dependent results have been obtained in some renormalizable field theories in low order perturbation theory. These results will be presented in the next lecture.

Note added in proof: A review of the field theoretic aspects of current algebra has been given recently by *C. A. Orzalesi* (University of Maryland-Technical Report No. 833).

References

1. *Streater, R. F.*, and *A. S. Wightman:* PCT, spin & statistics and all that. New York: W. A. Benjamin 1964.
2. *Schroer, B.*, and *P. Stichel:* Commun. math. Phys. **3**, 258 (1966).
3. *Ruelle, D.:* Helv. Phys. Acta **35**, 147 (1962). — *Araki, E.*, *K. Hepp*, and *D. Ruelle:* Helv. Phys. Acta **35**, 164 (1962).
4. *Steinmann, O.:* private communication.
5. *Schroer, B.*, and *P. Stichel:* Commun. math. Phys. **8**, 327 (1968).
6. *Källén, G.:* Helv. Phys. Acta **25**, 417 (1952). — *Lehmann, H.:* Nuovo Cimento **11**, 342 (1954).
7. *Schroer, B.:* Diplomarbeit, Hamburg 1958, unpublished. — *Jost, R.:* Lectures on field theory. Naples: 1959. — *Federbush, P. G.*, and *K. A. Johnson:* Phys. Rev. **120**, 1926 (1960).
8. *Goto, T.*, and *T. Imamura:* Progr. Theoret. Phys. (Kyoto) **14**, 396 (1955). — *Schwinger, J.:* Phys. Rev. Lett. **3**, 296 (1959). — *Okubo, S.:* Niovo Cimento **44**, 1015 (1966).

Professor Dr. *P. Stichel*
Physikalisches Staatsinstitut der Universität Hamburg
2000 Hamburg 50, Luruper Chaussee 149 and
Deutsches Elektronensynchrotron (DESY) Hamburg
2000 Hamburg 52, Notkestieg 1

Current Algebra and Renormalizable Field Theories

P. Stichel

Contents

1. Introduction

One of the basic ideas behind *Gell-Mann's* current algebra concept [1] is to give a formulation of symmetry breaking in strong interaction dynamics in terms of equal-time commutation relations (ETCR) between appropriate chosen currents. It is therefore a natural question to ask, whether current algebra can be realized in the case of an arbitrary symmetry breaking or only in very special cases. One possibility to investigate this problem is to study ETCR in renormalizable field theories in perturbation theory.

In this talk we will discuss renormalizable field theories describing the interaction of spin zero particles (π- and σ-mesons). First we investigate equal-time commutators (ETC) between axial charges in second order perturbation theory. This is the most relevant part of current algebra, because such successful relations like the *Adler-Weisberger* relation [2] are based on it.

Furthermore, results on ETCR of currents in 1st order perturbation theory will be presented.

In contradistinction to *Johnson* and *Low* [3], *Polkinghorne* [4], *Hamprecht* [5] and others we do not consider such ambiguous quantities as time-ordered products of currents, or introduce any cut-off. The different cut-off procedures used by these authors led them to conclude that non-canonical terms in ETCR are not well defined. But by calculating commutators of renormalized currents directly by means of renormalized field operators in perturbation theory one obtains in all cases, where the equal-time limit exists, unambiguous results.

2. Interactions and Currents

We consider renormalizable interactions between a pion-field $\vec{\pi}(x)$ and a σ-meson field $\sigma(x)$. All point interactions without derivatives which are at most quadrilinear in these fields are renormalizable. We restrict ourselves to isospin conserving interactions.

The following types of interactions will be considered explicitly (given in terms of unrenormalized quantities):

$$L_1 = g_{1u}\vec{\pi}_u^2\sigma_u\,; \qquad L_2 = g_{2u}\vec{\pi}_u^2\sigma_u^2\,,$$
$$L_{G-L} = g_u(\vec{\pi}_u^2 + \sigma_u^2)^2 + 2g_u\varkappa\sigma_u(\sigma_u^2 + \vec{\pi}_u^2) + g_u\varkappa^2\sigma_u^2\,. \tag{1}$$

L_{G-L} describes the mesonic part of Gell-Mann and Levy's σ-model [6] where now $\varkappa$ is a free parameter. The isovector-vector current is defined as usual*

$$j_{\mu i}^{(v)}(x) = Z_\pi \tfrac{1}{4}\varepsilon_{inm}\{\pi_n(x), \overleftrightarrow{\partial}_\mu \pi_m(x)\}\,. \tag{2}$$

ETCR's of the $j_{\mu,i}^{(v)}$ in the free field case (i.e. in zeroth order of perturbation theory) contain q-number gradient terms which are symmetric in the isospin indices [7]. Therefore, Gell-Mann's $SU(2)$ current algebra is satisfied only by the isospin antisymmetric part.

The most general isovector-axialvector current containing at most bilinear terms in the fields and fulfilling together with (2) Gell-Mann's $SU(2) \times SU(2)$ current algebra in the free field case** is given by [8]

$$j_{\mu,i}^{(A)}(x) = \tfrac{1}{2}\{\sigma(x), \overleftrightarrow{\partial}_\mu \pi_i(x)\} + c\,\partial_\mu \pi_i(x) \qquad + \text{counter terms}\,. \tag{3}$$

The counter terms in (3) serve to renormalize $j_\mu^{(A)}$ in higher order perturbation theory. It is generally believed that their number is finite [9]. For our purpose the knowledge of their explicit structure is not needed.

We note a particular property of the axial-vector current for the interaction L_{G-L}: If we put $c = \varkappa/2$ then the divergence of $j_{\mu,i}^{(A)}$ is proportional to the canonical π-field [6]

$$\partial^\mu j_{\mu,i}^{(a)}(x) \sim \pi_i(x)\,. \tag{4}$$

---

* Here and in the following all field operators are renormalized Heisenberg field operators.

** Again this is only true for combinations of current commutators being antisymmetric in the internal quantum numbers.

Throughout this paper we put the masses of π and σ equal ($=m$). This leads to $\partial^\mu j^{(A)}_{\mu,i}=0$ for $c=0$ in the free field case.

We compute field operators and with them our currents in perturbation theory by means of iteration solutions of renormalized Yang-Feldman equations:

$$\begin{aligned}\sigma(x)&=\sigma^{\text{in}}(x)+\int d^4x'\,\Delta_R(x-x')\,j_\sigma(x')\\ \pi_i(x)&=\pi_i^{\text{in}}(x)+\int d^4x'\,\Delta_R(x-x')\,j_i(x')\end{aligned}\tag{5}$$

where e.g. in the case of the $\vec{\pi}^2\,\sigma^2$-coupling the currents j_σ, j_π are given by

$$\begin{aligned}j_\sigma(x)&=gZ_1Z_\sigma^{-1}:\vec{\pi}^2(x)\,\sigma(x):+\delta m_\sigma^2\sigma(x)\,,\\ j_{\pi_i}(x)&=gZ_1Z_\pi^{-1}:\pi_i(x)\,\sigma^2(x):+\delta m_\pi^2\pi_i(x)\,.\end{aligned}\tag{6}$$

The renormalization constants are fixed by the conditions

$$\begin{aligned}&\langle 0|\pi_i(0)|1\rangle=\langle 0|\pi_i^{\text{in}}(0)|1\rangle=(2\pi)^{-3/2}\\ &\langle 0|\sigma(0)|1\rangle\;=\langle 0|\sigma^{m}(0)|1\rangle=(2\pi)^{-3/2}\\ &g\;=\;\text{four point function for fixed momenta}^{\star}\,.\end{aligned}$$

3. ETCR of Axial Charges in Second Order Perturbation Theory

We want to check ETCR between axial-charges which formally lock as follows

$$[Q_+^5(t), Q_-^5(t)]=2I_3\,.\tag{7}$$

Other isospin combinations of ETCR of the Q_i^5 may be obtained from (7) by means of an isospin rotation, i.e. they are valid if and only if (7) is valid.

3.1 One Particle Matrix Elements

Now we consider the matrix element of (7) between positive charged one-pion states. For this matrix element we use the rigorous formulation of ETCR given in the last lecture. By means of the usual manipulations which have been proved in [10] on the level of general quantum field theory, we arrive at the following equivalent relation

$$G_A^2+\frac{(2\pi)^2}{4p_0}\lim_{T\to 0}\int dq_0\tilde{f}_T(q_0)\frac{M(q_0,\vec{0})_p}{q_0^2}=1\tag{8}$$

$\star$ For example one may put all momenta equal to zero.

where the axial-vector coupling constant G_A and the matrix element $M(q)_p$ are defined by

$$\langle \pi_+(p)| j^{(A)}_{\mu+}(0)|\sigma(p)\rangle = i\sqrt{2}\, G_A \frac{2p_\mu}{(2\pi)^3} \tag{9}$$

and

$$M(q)_p = \int d^4x\, e^{iqx} \langle \pi_+(p)|\,[D_+(x/2), D_-(-x/2)]\,|\pi_+(p)\rangle \tag{10}$$

respectively, with

$$D_i(x) = \partial^\mu j^{(A)}_{\mu i}(x)\,.$$

$\{\tilde{f}_T\}$ denotes the Fourier-transform of the sequence of time smearing functions $\{f_T\}$ ($\{\tilde{f}_T\} \to 1$).
In second order perturbation theory (8) is equivalent to

$$2G_A^{(2)} + \frac{(2\pi)^2}{4p_0} \lim_{T\to 0} \int dq_0 \tilde{f}_T(q_0) \frac{M^{(2)}(q_0,\vec{0})_p}{q_0^2} = 0\,. \tag{11}$$

Due to the use of a renormalized axial-vector current $G_A^{(2)}$ is always finite. On the other hand the counter terms in (3) are only fixed up to a finite contribution. These finite contributions, i.e. the numerical value of $G_A^{(2)}$ e.g. may be fixed by (11) if the equal-time limit exists.

Therefore, we conclude: (8) is fulfilled in second order perturbation theory if and only if

$$\frac{(2\pi)^2}{4p_0} \lim_{T\to 0} \int dq_0 \tilde{f}_T(q_0) \frac{M^{(2)}(q_0,\vec{0})_p}{q_0^2} \tag{12}$$

exists as an p_0 independent number which is independent of the sequence $\{\tilde{f}_T\}$*.

Now we investigate (12) in case of the different interactions mentioned above.

$\vec{\pi}^2\sigma^2$-Coupling

In the following we will show that (12) diverges for the following sequence of symmetric test functions

$$\tilde{f}_T(q_0) = e^{-T^2 q_0^2}\,. \tag{13}$$

a) $c=0$.

In this case $D_i^{(0)}=0$ i.e. for the computation of $M^{(2)}(q)_p$ we need, according to (10), only $D_\pm^{(1)}$. In first order counter terms appear neither in the field equation nor in the currents. We obtain [8]

$$D_i^{(1)}(x) = -\,:\vec{\pi}^{(0)^2}(x)\,\pi_i^{(0)}(x)\,\sigma^{(0)}(x): + :\pi_i^{(0)}(x)\,\sigma^{(0)^3}(x):. \tag{14}$$

* For the choice of sequences $\{f_T\}$ compare the last lecture.

Due to the use of symmetric test functions (13) we only need the isospin-antisymmetric part of (10) $M_a^{(2)}$. By means of straightforward computations one obtains from (14)

$$M_a^{(2)}(q)_p = 16(2\pi)^{-2} \int d\varkappa^2\, \varphi(\varkappa^2)\, \varepsilon(q+p)\, \delta((q+p)^2 - \varkappa^2) + (q \to -q)\,. \quad (15)$$

The number 16 in front of (15) just counts the different possibilities of triple contractions in $[D_{[+]}^{(1)}(x), D_{[-]}^{(1)}(y)]$. The spectral function $\varphi(\varkappa^2)$ is defined by

$$\Delta_m^{(+)^3}(x) = -\int d\varkappa^2\, \varphi(\varkappa^2)\, \Delta_\varkappa^{(+)}(x)\,. \quad (16)$$

Explicitely, we have

$$\varphi(\varkappa^2) = \int d\varkappa'^2\, \varrho_{mm}(\varkappa'^2)\, \varrho_{m\varkappa'}(\varkappa^2)$$

with

$$\varrho_{ab}(c^2) = \frac{(2\pi)^{-2}}{4c} \sqrt{a^2+b^2+c^2-2ab-2ac-2bc}\; \Theta(c^2-(a+b)^2)$$

Therefore, $\varphi(\varkappa^2)$ has the asymptotic behaviour

$$\varphi(\varkappa^2) \xrightarrow[\varkappa^2 \to \infty]{} o(\varkappa^2)\,. \quad (17)$$

By using for $\{\tilde{f}_T\}$ the sequence (13) we obtain finally for (12) the result

$$16 \lim_{T \to 0} \int d\varkappa^2 \frac{\varphi(\varkappa^2)}{(\varkappa^2 - m^2)^2}\, e^{T^2(p_0^2 + \vec{p}^2 + \varkappa^2)} \operatorname{Cosh}(2T^2 p_0 \sqrt{\vec{p}^2 + \varkappa^2}) \quad (18)$$

which due to (17) diverges logarithmically*.

b) $c \neq 0$.

The result (18) is not altered, because

α) in the computation of $M^{(2)}$ appear no cross terms between the two contributions to $j_\mu^{(A)}$ in (3);

β) the truncated ETC of the $\dot{\pi}_i$, i.e.

$$\lim_{T \to 0} \int dx_0 f_T(x_0)\, [\dot{\pi}_i(x/2), \dot{\pi}_k(-x/2)]_{\mathrm{tr}}^{(2)} \quad (19)$$

vanishes in agreement with the canonical commutation relations of field operators. This may be seen by a straightforward but lengthy computation of (19) with the aid of methods developed in [8].

We note that the divergent term in (18) is proportional to the wave function renormalization constant of the π-respectively σ-field in second order.

---

* Such a situation has been found recently by *Katz* and *Langerholc* [13] in the case of quantum electrodynamics too.

$\vec{\pi}^2\sigma$-Coupling

a) $c=0$.

In this case the computation of $M_a^{(2)}$ only contains double contractions. The final result has the same form as above if we make the substitution

$$16\,\varphi(\varkappa^2)\to\tfrac{18}{4}\varrho_{mm}(\varkappa^2)\,,$$

i.e. due to $\varrho_{mm}(\varkappa^2)\xrightarrow[\varkappa^2\to\infty]{}o(1)$ we get a finite p_0 and sequence independent number for (12). This is not surprising as again the result is proportional to the wave function renormalization constant and the latter is known to be finite.

b) $c\neq 0$.

We may repeat the discussion given in case of the $\vec{\pi}^2\,\sigma^2$-coupling.

Gell-Mann-Levy-Type Coupling

Due to $D_i(x)\sim\pi_i(x)$ expression (12) exists as a p_0 and sequence independent number.

Proof: The high energy behaviour of $M^{(2)}(q)_p$ is given by the quadrilinear terms in the interaction. By means of methods developed in [8] one finds that the Dyson spectral function $\varphi(u,s)_p$ in the Dyson representation for $M^{(2)}(q)_p$

$$M^{(2)}(q)_p=\int d^4u\int ds\,\varepsilon(q-u)\,\delta((q-u)^2-s)\,\varphi(u,s)_p$$

has the asymptotic behavior

$$\varphi(u,s)_p\xrightarrow[s\to\infty]{}o(s^{-2})\,. \tag{20}$$

Therefore, according to [10] (12) exists and is given by

$$-\frac{(2\pi)^2}{4p_0}\int d^4u\int ds\,\frac{2u_0\,\varphi(u,s)_p}{(u^2-s)^2}\,. \tag{21}$$

As φ is an invariant function (21) turns out to be p_0 independent.

The asymptotic behaviour (20) is also responsible for the validity of the canonical commutation relations of field operators in second order.

3.2 Many Particle Matrix Elements

An analysis of the structure of the quadrilinear terms (with respect to the free fields) in the commutator

$$\int d^3x\int d^3y\,[j_{0,i}^{(A)}(x),j_{0,k}^{(A)}(y)]$$

in the case of the $\vec{\pi}^2\sigma$-coupling and the Gell-Mann-Levy-type coupling respectively leads to the result that in the equal-time limit only "tree graphs", i.e. terms containing only one contraction contribute. According to the kind of arguments to be presented for one-contraction terms in chapter 4 these terms only lead to normal, i.e., "canonical" results.

3.3 Further Remarks on Currents and Interactions

In the case of the $\vec{\pi}^2\sigma$-coupling and the Gell-Mann-Levy-type coupling respectively we conclude that ETCR of axial charges are in agreement with Gell-Mann's $SU(2) \times SU(2)$ current algebra in second order perturbation theory.

But from the point of view of general quantum field theory the $\vec{\pi}^2\sigma$-coupling is not admissible, because it forbids due to the nonpositive definiteness of the Hamilton operator a state of lowest energy, i.e. no vacuum state exists [11]. This statement destroys also the basis for the usual perturbation theoretical treatment of this interaction.

In the case of the $\vec{\pi}^2\sigma^2$-coupling the equal-time limit of axialcharge commutators does not exist. But this result may depend on the special form of our axial-vector current for which we used the most general form which is at most bilinear in the fields and satisfies the $SU(2) \times SU(2)$ algebra in the free field case. From our experience with the Gell-Mann-Levy-type coupling it is suggested to look for a more general current whose divergence is proportional to the canonical pion field. It is easily seen that such a current cannot exist in the case of a $\vec{\pi}^2\sigma^2$-coupling: Suppose we have $\partial^u j^{(A)}_{\mu,i}(x) = c\pi_i(x)$ ($c \neq 0$). Then we take the matrixelement of this relation between a one-pion and a two-σ-state

$$\langle \sigma_1 \sigma_2 | \partial^\mu j^{(A)}_{\mu,i}(0) | \pi \rangle = c \langle \sigma_1 \sigma_2 | \pi_i(0) | \pi \rangle . \tag{22}$$

According to *Adler* [12] only graphs of the following structure contribute to the l.h.s. of (22) in the limit $p_1 + p_2 - p_\pi \to 0$.

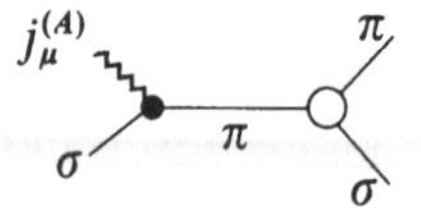

But the $\pi - \pi - \sigma$ vertex vanishes identical in our case, as the $\pi^2\sigma^2$-theory is invariant with respect to the charge conjugation $\pi_m \to \pi_{-m}\ \sigma \to -\sigma$. On the other hand the r.h.s. of (22) does not vanish in this limit, as e.g.

$$\langle \sigma_1 \sigma_2 | \pi_i(0) | \pi \rangle \underset{p_1 + p_2 - p_\pi \to o}{\propto} \frac{g}{m^2}$$

in first order perturbation theory.

This is a contradiction. Nevertheless, it is still an open question whether a more general axial-vector current exists which satisfies current algebra in second order perturbation theory in the case of the $\pi^2\sigma^2$-coupling.

4. ETCR of Currents in First Order Perturbation Theory

We consider the $\vec{\pi}^2\sigma^2$-coupling as a specific example [8].

The first order commutators:

$$[j^{\alpha}_{\mu i}(x), j^{\beta}_{\nu k}(y)]^{(1)} = [j^{\alpha(0)}_{\mu i}(x), j^{\beta(1)}_{\nu k}(y)] + [j^{\alpha(1)}_{\mu i}(x), j^{\beta(0)}_{\nu k}(y)]$$

contain terms quadrilinear, trilinear and bilinear in the free fields arising from single resp. double contractions.

The single contraction terms do not lead to any anomalous result. Either two fields at the points x and y are contracted with each other leading to a $\Delta(x-y)$ function whose time derivative gives at equal times normal, finite terms (those terms also contain gradient terms, as they are of the same canonical structure as the zeroth order commutators), or $\underbrace{x\ x'}$ resp. $\underbrace{y\ x'}$ contractions take place whose sum leads to the local chain:

$$\int d^4x'\, P_{\mu\nu}(\partial^x, \partial^y)\,(\Delta_A(x-x')\,\Delta_A(x'-y) - \Delta_R(x-x')\,\Delta_R(x'-y)) : \sigma^{(0)2}(x') : \tag{23}$$

where $P_{\mu\nu}(\partial^x, \partial^y)$ is a second order polynomial in the derivatives. By means of contour integration in momentum space it is very easy to see that (23) vanishes for $x_0 = y_0$.

Therefore, we concentrate our effort on the bilinear terms which follow from double contractions. By means of straight-forward computation we obtain:

$$[j_{\mu\alpha}{}^{(x)}\, j_{\nu\beta}{}^{(y)}]^{(1)}_{\text{bilinear}} = \int d^4x'\, I_{\mu\nu}(x, y, x')\, K_{\alpha\beta}(x') \tag{24}$$

where $K_{\alpha\beta} = K_{\beta\alpha}$ is an expression bilinear in the free field operators at the point x'.

$I_{\mu\nu}(x, y, x')$ is defined as follows:

$$\begin{aligned} I_{\mu\nu}(x, y, x') = -i\{&-[\Delta^{(1)}(y-x')\,\Delta_A(x'-x) + \Delta_R(y-x')\,\Delta^{(1)}(x'-x)] \\ &\overleftrightarrow{\partial}^x_\mu\, \overleftrightarrow{\partial}^y_\nu\, \Delta(x-y) + [\Delta_A(y-x')\,\Delta_A(x'-x) - \Delta_R(y-x')\,\Delta_R(x'-x)] \\ &\overleftrightarrow{\partial}^x_\mu\, \overleftrightarrow{\partial}^y_\nu\, \Delta^{(1)}(x-y)\}\,. \end{aligned} \tag{25}$$

As the vector current is conserved, $I_{\mu\nu}$ is divergenceless, i.e.:

$$\partial^\mu_x I_{\mu\nu}(x, y, x') = \partial^\nu_y I_{\mu\nu}(x, y, x') = 0\,.$$

It is amusing to note that according to equation (24) our axialvector current is effectively conserved in the bilinear part of the commutators.

Because (25) corresponds to the triangle graph (to be understood as a double graph for the retarded resp. advanced current commutator)

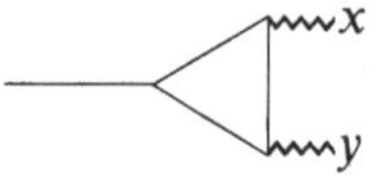

it is not surprising that $I_{\mu\nu}$ satisfies a Dyson representation of the vertical type, i.e. in momentum space we have

$$\tilde{I}_{\mu\nu}(q,\Delta)=\int d^4x'\, e^{-i\Delta x'}\int d^4\xi\, e^{iq\xi}\, I_{\mu\nu}(\xi/2,-\xi/2,x'))$$

$$\tilde{I}_{\mu\nu}(q,\Delta)=\int_{-1/2}^{+1/2} dx\int_0^\infty ds\,\varepsilon(q_0-x\Delta_0)\,\delta((q-x\Delta)^2-s)\,. \qquad (26)$$

$$\cdot\sum_{i=1}^{5} P^i_{\mu\nu}(q,\Delta)\,\Psi^i(\Delta^2,s,x)$$

where the $P^i_{\mu\nu}(q,\Delta)$ are the five possible standard tensors $g_{\mu\nu}$, $q_\mu q_{\nu_i}$, $q_\mu \Delta_\nu$, $q_\nu \Delta_\mu$, $\Delta_\mu \Delta_\nu$. By means of explicit expressions for the Ψ^i given in [8] it is relatively easy to compute the ETC.

The structure of the results for the ETC's are independent of the value of Δ [8]. Therefore, we will discuss explicitely in the following only the particular case $\Delta=0$. In this case, the vertical Dyson-representation (26) reduces to the Källén-Lehmann representation. We obtain

$$\sum_i P^i_{\mu\nu}\,\Psi^i\Big|_{\Delta=0}\propto\left(g_{\mu\nu}-\frac{q_\mu q_\nu}{s}\right)\frac{3s+\mu^2}{s}\sqrt{1-\frac{4\mu^2}{s}}\;\Theta(s-4\mu^2)\,,$$

i.e. according to the arguments given for the calculation of ETC of VEV of currents in the last lecture it follows in the equal time limit:

a) zero in the case of time-time resp. space-space components,

b) non-existence of the ETC in the case of time-space components (logarithmic divergence if we use the particular sequence $\{f_T\}$ introduced in chapter 3.1).

The same results are valid for arbitrary Δ and also for the other two types of interactions [8].

Our methods have been applied recently by *Usyukina* [13] to first order perturbation calculations of ETCR within the *Johnson-Low* model [3]. The results are similar to ours.

5. Concluding Remarks

It would be interesting to know whether q-number gradient terms which are antisymmetric in the internal quantum numbers of currents exist in higher than first order perturbation theory.

From our results we conclude that at present the only type of renormalizable interaction for spin zero fields which supports Gell-Mann's $SU(2) \times SU(2)$ current algebra hypothesis (for that part which is antisymmetric in the internal quantum numbers) is the Gell-Mann-Levy interaction.

This result already strongly suggests that the requirement of ETCR at least for generalized charges is a dynamical requirement.

References

1. *Gell-Mann, M.:* Phys. Rev. **125**, 1067 (1962).
2. *Adler, A.:* Phys. Rev. Letters **14**, 1051 (1965); — *Weisberger, W. I.:* Phys. Rev. Letters **14**, 1047 (1965).
3. *Johnson, K.*, and *F. E. Low:* Progr. Theor. Phys. Suppl. **37**, **38**, 74 (1966).
4. *Polkinghorne, J. C.:* Nuovo Cimento **52** A, 351 (1967).
5. *Hamprecht, B.:* Nuovo Cimento **47** A, 770 (1967) and **50** A, 449 (1967).
6. *Gell-Mann, M.*, and *M. Levy:* Nuovo Cimento **16**, 705 (1960).
7. *Adler, S. L.*, and *C. G. Callan:* CERN, TH 587 (1965).
8. *Schroer, B.*, and *P. Stichel:* Commun. Math. Phys. **8**, 327 (1968).
9. *Wilson, K.:* unpublished manuscript.
10. *Schroer, B.*, and *P. Stichel:* Comm. Math. Phys. **3**, 258 (1966).
11. *Baym, G.:* Phys. Rev. **117**, 886 (1960).
12. *Adler, S. L.:* Phys. Rev. **139**, B 1638 (1965).
13. *Katz, J.*, and *J. Langerholc:* DESY 68/53.
14. *Usyukina, N.:* Dubna preprint P2-3661 (1968).

Prof. Dr. *P. Stichel*
Physikalisches Staatsinstitut der Universität Hamburg
2000 Hamburg 50, Luruper Chaussee 149
and Deutsches Elektronensynchrotron (DESY) Hamburg
2000 Hamburg 52, Notkestieg 1

Introduction to Current Algebra

P. STICHEL

Contents

1. Introduction

Some of the progress made during the last years in the theory of elementary particles is connected with the concept of current algebra. I am trying to give in this talk a short review of the content of current algebra, its main results and its role within present elementary particle physics.

The content of current algebra may be characterized as follows: Current algebra is the equal-time commutator algebra of hadronic currents appearing in weak and electromagnetic interactions. These equal time commutation relations are of the following kind

$$[j_\mu^\alpha(x), j_\nu^\beta(y)]_{x_0=y_0} = g_{\mu\nu\varrho}^{\alpha\beta\gamma} j_\varrho^\gamma(x)\,\delta(\boldsymbol{x}-\boldsymbol{y})$$

thereby α, β, γ are internal quantum numbers [$SU(3)$-index, axial- resp. vector current] and the $g_{\mu\nu\varrho}^{\alpha\beta\gamma}$ are structure constants of either an exact or an approximate symmetry group.

2. The Role of Current Algebra within Present Elementary Particle Physics

In contrast to the situation in atomic or low energy nuclear physics we don't know the basic dynamical equations in elementary particle physics, i.e. we don't know the equations determining, at least in principle, everything. What we call at present theory of elementary particles are only some small pieces consisting mainly of

a) the general framework of axiomatic field theory, supplemented by certain symmetry principles. This framework is of such a generality that it may contain several possible theories of elementary particles.

b) Some ideas on the analytic properties of the S-matrix as a function of the kinematical variables. Part of this S-matrix analyticity may be proven within axiomatics, the remainder are postulates based on our experience in potential theory or on some results obtained in the laboratory of Feynman graphs.

c) Dynamical models. For the purpose of illustration I only mention some well known slogans: Peripheral model, Regge poles, current-current coupling in weak interactions etc. We hope that such models are good approximations to the yet unknown theory of elementary particles in certain situations.

What is the predictive power of these pieces of the theory? According to a) and b) we may predict relations between different measurable quantities (e.g. dispersion relations, i.e. sum rules). On the other hand dynamical models are often able to make quantitative predictions if the numerical values of certain coupling constants are known.

Where is the place for current algebra in this scheme? It is very difficult to integrate it into either a), b) or c). Current algebra is a piece of the dynamics, it is a dynamical principle within the framework of general quantum field theory, in particular it is a general formulation of

1. symmetry breaking in strong interaction dynamics,

2. the universal coupling of hadron currents in weak interactions.

The predictive power of current algebra is of different nature: On the one hand it leads to sum rules but on the other hand it leads also to quantitative predictions of transition amplitudes at threshold (low energy theorems).

The direct connection between commutation relations and sum rules is known since a long time. The general idea looks as follows:

From a commutator between observables

$$[A, B] = C$$

we take matrix elements, introduce a complete set of intermediate states $1 = \sum_n |n\rangle \langle n|$ and obtain in this way the sum rule

$$\sum_n \{\langle b|A|n\rangle \langle n|B|a\rangle - \langle b|B|n\rangle \langle n|A|a\rangle\} = \langle b|C|a\rangle .$$

An almost classical example of such a sum rule is the dipole sum rule

$$2M \sum_n (E_n - E_a)| Q_{1,0}(a, n)|^2 = \left(\frac{3}{4\pi}\right)\left(\frac{NZ}{A}\right) e^2$$

with

$$Q_{1,0}(a,n) \equiv : e\left(\frac{3}{4\pi}\right)^{1/2} \sum_{k=1}^{Z} \langle n|z_k|a\rangle$$

which is based on *Heisenberg*'s commutation relations between coordinate and momentum operators $[z_k, p_{k'z}] = i\delta_{kk'}$.

3. $SU(3) \times SU(3)$ Algebra of Currents

We are now going to discuss in detail the chiral $SU(3) \times SU(3)$ algebra of currents postulated already in 1962 by Gell-Mann. We start with the chiral $SU(2) \times SU(2)$ algebra of charges resp. generalized charges.

The interaction between hadrons is invariant with respect to the symmetry group $SU(2)_I$ leading to the isospin Q as a conserved quantity. The commutator algebra of the Q_i is the representation of the $SU(2)_I$ Lie algebra in Hilbert space $\mathfrak{H}$: $[Q_i, Q_k] = i\varepsilon_{ikr} Q_r$. The isospin operators are connected in a local field theory with isospin currents $j^{(v)}_{\mu,i}$ as follows

$$Q_i = \int d^3x\, j^{(v)}_{0,i}(x).$$

Conservation of isospin, i.e. $[Q_i, H] = 0$ leads to a conservation law for isospin currents: $\partial^\mu j^{(v)}_{\mu,i} = 0$. On the other hand isospin currents appear in the electromagnetic resp. weak interactions. Therefore, they are observable quantities. We have

a) $j_\mu^{\text{el,Hadron}} = e(j_\mu^{(s)} + j^{(v)}_{\mu,3})$,

b) $$H^{\Delta S=0,L}_{\text{weak}}(x) = \frac{g\cos\theta}{\sqrt{2}}\,(j^{(v)}_{\mu,1+i2}(x) + j^{(A)}_{\mu,1+i2}(x))\,\overline{\psi}_e(x)\gamma_\mu(1-\gamma_5)\psi_\nu(x) + \text{h.c.}$$

where θ is *Cabibbo*'s angle.

The identification of hadronic vector currents in H_{weak} with the isospin currents $j^{(v)}_{\mu,i}$ fixes their scale in such a way, that they are coupled to other currents with the same strength as the lepton currents (principle of a universal current-current coupling).

Besides vector currents $j^{(v)}_{\mu,i}$ also axial vector currents $j^{(A)}_{\mu,i}$, transforming like components of an isovector, are present in weak interactions. Therefore, both kinds of currents satisfy the commutation relations

$$[Q_i, j_{\mu,k}(x)] = i\varepsilon_{ikr} j_{\mu,r}(x).$$

Now we have to formulate the principle of a universal coupling of the $j^{(A)}_{\mu,i}$ in weak interactions. For this we proceed as follows:

In analogy to isospin operators, also called vector charges, we may, at least formally, define axial charges*

$$\text{“}Q_i^5\text{”} \equiv : \int d^3x\, j_{0,i}^{(A)}(x) .$$

The axial vector currents can be conserved only if either

a) we have an exact symmetry with vector and axial charges as generators (chiral $SU(2) \times SU(2)$). Then all hadrons have to occur as parity dubletts

or b) we have a spontaneously broken symmetry, i.e. zero rest mass particles with the quantum numbers of the pions must exist as Goldstone particles.

Neither of the two possibilities is realized in nature, but, due to the smallness of the pion mass compared to other hadron masses, possibility b) may be considered as approximately valid.

According to Gell-Mann we formulate this symmetry breaking as follows: The time dependent axial charges satisfy in form of equal time commutation relations (ETCR) the same algebra like the time independent charges in the limit of a spontaneously broken symmetry, i.e.

$$[Q_i^5(x_0), Q_k^5(x_0)] = i\varepsilon_{ikr}\lambda Q_r .$$

The real constant λ will be fixed by the principle of a universal current coupling in weak interactions: The Q_i^5 should obey the same ETCR as the integrated axial lepton currents. In this way we obtain $\lambda = 1$.

These postulates are nontrivial in two respects:

1. The equal time limit of the commutation relations between axial charges exists,

2. the equal time commutation relations have the particular form given above.

Gell-Mann's recipe may be generalized in case of a higher symmetry like $SU(3)$ e.g.

It is well known from experimental facts that $SU(3)$ is broken by some semistrong interactions. According to Gell-Mann we may again formulate this symmetry breaking as equal time commutation relations between the corresponding $SU(3)$ charges.

The $SU(3)$ charges $Q_i = \int d^3x\, j_{0,i}^{(v)}(x)$, $(i = 1$—$8)$, may be obtained in analogy to the $SU(2)$ case by space integration of hadron currents occuring in weak resp. electromagnetic interactions. For this we need in addition to the Hamiltonians considered above $H_{\text{weak}}^{|\Delta S|=1,L}$. We have

$$H_{\text{weak}}^{|\Delta S|=1,L} = \frac{g}{\sqrt{2}} \sin\theta \{j_{\mu,4+i5}^{(v)}(x) + j_{\mu,4+i5}^{(A)}(x)\}\, \overline{\psi}_e(x)\gamma_\mu(1-\gamma_5)\psi_\nu(x) + \text{h.c.}$$

with $\overset{(A)}{j_{\mu,i}^{(v)}}$, $i = 1 \ldots 8$ forming an octet of currents in the limit of exact $SU(3)$.

* The rigorous definition of such generalized charges will be discussed in another lecture.

We identify from the octet of vector resp. axial vector currents 6 resp. 4 in the Hamiltonians for weak and electromagnetic interactions.

In analogy to $SU(3)$ vector charges we may introduce (again formally) $SU(3)$ axial charges

$$Q_i^5 \equiv: \int d^3x\, j_{0,i}^{(A)}(x)\,, \qquad i = 1 \ldots 8\,.$$

The total number of these 16 charges would constitute the algebra of chiral $SU(3) \times SU(3)$ in the symmetry limit. The fact of symmetry breaking is taken into account again by requiring the algebra of $SU(3) \times SU(3)$ as commutator algebra of these time dependent charges at equal times:

$$\begin{aligned}
[Q_i(x_0), Q_k(x_0)] &= i f_{ikr} Q_r(x_0)\,,\\
[Q_i(x_0), Q_k^5(x_0)] &= i f_{ikr} Q_r^5(x_0)\,,\\
[Q_i^5(x_0), Q_k^5(x_0)] &= i f_{ikr} Q_r(x_0)\,.
\end{aligned}$$

In this way the universal coupling of $|S| = 1$ hadron currents in weak interactions has been formulated too.

What is known about the equal time commutation relations of current components? In a local field theory the algebra of charges only fixes the term proportional to the three-dimensional δ-function in case of current commutation relations. Consider as an example ETCR between time components of currents. We have in case of $SU(3)$ vector currents e.g.

$$[j_{0,i}^{(v)}(x), j_{0,k}^{(v)}(y)]_{x_0 = y_0} = i f_{ikr} j_{0,r}^{(v)}(x)\, \delta(\boldsymbol{x} - \boldsymbol{y}) + \cdots\,.$$

The dots mean that additional terms may occur which are proportional to derivatives of δ-functions, i.e. vanish after space integration.

We have an analogous result for the ETCR between time- and space components of currents. From the covariance requirement

$$[Q_i, j_{\mu,k}(x)] = i f_{ikr} j_{\mu,r}(x)$$

in the case of conserved charges (which may be generalized to an ETCR for non-conserved charges) it follows that only the term proportional to the three-dimensional δ-function is fixed in the ETCR between time and space component of currents. Gell-Mann made the hypothesis that no gradient terms appear in ETCR between current components. But on the other hand it is known that such gradient terms must occur in the vacuum expectation value of ETCR of currents. This statement follows alone from Lorentz invariance and the spectrum conditions and was first made in 1955 by Imamura and Goto. It was rediscovered in 1959 by Schwinger. The proof of this statement will be given in another lecture. Therefore, before applying Gell-Mann's hypothesis we must subtract the vacuum expectation values, i.e. we have to consider the truncated commutators. Then Gell-Mann's hypothesis looks as follows:

$$[j^{(v)}_{0,i}(x), j^{(v)}_{\mu,k}(y)]^{\text{tr.}}_{x_0=y_0} = i f_{ikr} j^{(v)}_{\mu,r}(x)\, \delta(\boldsymbol{x}-\boldsymbol{y})$$
$$[j^{(v)}_{0,i}(x), j^{(A)}_{\mu,k}(y)]^{\text{tr.}}_{x_0=y_0} = i f_{ikr} j^{(A)}_{\mu,r}(x)\, \delta(\boldsymbol{x}-\boldsymbol{y})$$
$$[j^{(A)}_{0,i}(x), j^{(v)}_{\mu,k}(y)]^{\text{tr.}}_{x_0=y_0} = i f_{ikr} j^{(A)}_{\mu,r}(x)\, \delta(\boldsymbol{x}-\boldsymbol{y})$$
$$[j^{(A)}_{0,i}(x), j^{(A)}_{\mu,k}(y)]^{\text{tr.}}_{x_0=y_0} = i f_{ikr} j^{(v)}_{\mu,r}(x)\, \delta(\boldsymbol{x}-\boldsymbol{y})$$

with

$$[A, B]^{\text{tr.}} \equiv : [A, B] - \langle 0|[A, B]|0\rangle \,.$$

We may now ask the question whether there exist field theoretic models where this $SU(3)\times SU(3)$ current algebra is valid. It is valid in the case of currents built up by free, i.e. non-interacting quark fields

$$\left.\begin{aligned} j^{(v)}_{\mu,i}(x) &= : \bar{\psi}(x)\gamma_\mu \lambda_i \psi(x) : \\ j^{(A)}_{\mu,i}(x) &= : \bar{\psi}(x)\gamma_5 \gamma_\mu \lambda_i \psi(x) : \end{aligned}\right. \qquad i = 1 \ldots 8 \,.$$

In case of interaction we may investigate this question in perturbation theory for renormalizable field theories. Let me just mention the result, details will be given in another lecture:

1. ETCR between generalized charges strongly depend on the kind of interaction. There are models, where even the equal time limit of commutators does not exist. This means essentially that the assumption of ETCR between generalized charges is a dynamical assumption.

2. In most of the models ETCR between currents contain q-number gradient terms. But they have been found only in that combination of current commutators which is symmetric in the internal quantum numbers.

In the following we will discuss characteristic examples concerning the application of current algebra. Let us start with sum rules.

4. The Adler-Weisberger Relation

The only one example where a sum rule following from current algebra may be compared with experimental numbers without introducing drastic assumptions is the Adler-Weisberger relation. This relation expresses the renormalization of the axial-vector coupling constant, i.e. the ratio

$$\frac{g_A}{g_v} \equiv G_A$$

in terms of total pion-nucleon cross sections.

We start with the ETCR between the + and − components of axial charges

$$[Q_+^5(x_0), Q_-^5(x_0)] = 2Q_3 .$$

Now we take the matrix element of this ETCR between one-proton states of equal momentum and obtain after some manipulations

$$1 = G_A^2\left(1-\frac{m^2}{p_0^2}\right)+\frac{(2\pi)^2}{2p_0}\int \mathrm{d}q_0\,\frac{M(q_0,\mathbf{0})_p}{q_0^2}$$

$$M(q)_p \equiv \int \mathrm{d}^4x\, \mathrm{e}^{iqx}\, \langle p|[D_+(x/2), D_-(-x/2)]\,|p\rangle \equiv M\{(q+p)^2, q^2\} .$$

In this form our sum rule is not yet comparable with experiment, because $M(s, q^2)$ which is proportional to $\operatorname{Im} T_{\pi^- p}(s, q^2)$ will be integrated over the pion mass $\sqrt{q^2}$.

The dependence of the r.h.s. of this sum rule on p_0 is only an apparent one. Under relatively weak conditions it may be shown that there is no p_0 dependence. Then we may evaluate the r.h.s. in the limit $p_0 \to \infty$. We assume that this limit may be interchanged with the q_0- integration. This is a non-trivial assumption which has nothing to do with current algebra. It is equivalent to the assumption that the retarded commutator matrix element corresponding to M obeys an unsubtracted dispersion relation in s for fixed $q^2 = 0$. We know by comparison with experimental cross sections that this is the case for physical pions, i.e. for $q^2 = \mu^2$.

In this way we obtain

$$1 = G_A^2 + (2\pi)^2 \int \mathrm{d}s\, \frac{M(s,0)}{(s-m^2)^2} .$$

We now use the fact that the divergence of the axial vector current is a possible interpolating field operator for pions

$$D_\pm(x) = c\,\phi_{\pi^\pm}(x), \qquad \frac{c}{\mu^2} = \frac{\sqrt{2}\, m G_A}{g_{\pi NN} K_{\pi NN}(0)} .$$

With this we obtain the Adler-Weisberger relation in its final form

$$1 = G_A^2 + \frac{1}{\pi}\frac{c^2}{\mu^4}\int\limits_{(m+\mu)^2}^{\infty} \mathrm{d}s\, \frac{\sigma_{\pi^- p}^{\mathrm{tot}}(s,0) - \sigma_{\pi^+ p}^{\mathrm{tot}}(s,0)}{s-m^2} .$$

Before using experimental cross sections for evaluating the r.h.s. of this sum rule we must continue them from the physical pion mass μ to the unphysical mass 0. We assume (PCAC assumption)

$$\sigma_{\pi p}^{\mathrm{tot}}(s,0)/K_{\pi NN}^2(0) \cong \sigma_{\pi p}^{\mathrm{tot}}(s,\mu^2) .$$

Finally we obtain by means of experimental cross sections from this sum rule the numerical value $G_A = 1.16$ which has to be compared with the experimental value 1.23 ± 0.02. By taking into consideration the additional assumptions having made this agreement is surprising. This success encourages us to make equivalent assumptions also for other applications of current algebra.

5. General Sum Rule à la Fubini

Fubini has given a general recipe for obtaining a sum rule from a given current commutation relation:

We start with the commutator

$$[j_{0,i}(x), j_{\mu,k}(y)]^{\text{tr.}}_{x_0=y_0} = i f_{ikr} j_{\mu,r}(x)\,\delta(\boldsymbol{x}-\boldsymbol{y})$$

and define the following tensors

$$T_{\mu\nu} \equiv i \int d^4x\, e^{iqx}\,\theta(x_0)\,\langle p_2|\,[j_{\mu,i}(x/2), j_{\nu,k}(-x/2)]\,|p_1\rangle$$

and

$$t_{\mu\nu} \equiv \int d^4x\, e^{iqx}\,\langle p_2|\,[j_{\mu,i}(x/2), j_{\nu,k}(-x/2)]\,|p_1\rangle$$

With the abbreviations

$$p \equiv \frac{p_1+p_2}{2}, \quad \Delta \equiv p_2 - p_1, \quad q_i \equiv q \pm \Delta/2\,,$$
$$s \equiv (q_i + p_i)^2, \quad t \equiv \Delta^2$$

we may introduce the following covariant decomposition of $T_{\mu\nu}$ resp. $t_{\mu\nu}$ in case of external particles with spin zero

$$T_{\mu\nu} = A(s,t,q_i^2)\,p_\mu p_\nu + \cdots\,,$$
$$t_{\mu\nu} = a(s,t,q_i^2)\,p_\mu p_\nu + \cdots\,.$$

An essential additional assumption is the hypothesis of unsubtracted dispersion relations in s for all of the invariant functions occuring in this decomposition

$$A(s,t,q_i^2) = \frac{1}{2\pi}\int ds'\,\frac{a(s',t,q_i^2)}{s'-s-i\varepsilon} \quad \text{etc.}$$

This leads finally to the sum rule

$$\frac{1}{2\pi}\int ds'\,a(s',t,q_i^2) = 2i f_{ikr} F(t)$$

whereby the formfactor $F(t)$ is defined as follows

$$\langle p_2|j_{\mu,r}(0)\,p_1\rangle = 2p_\mu F(t) + \Delta_\mu G(t)\,.$$

The Adler-Weisberger relation is a particular example of this general sum rule: take the commutator between axial vector currents and put in the sum rule $t = q_i^2 = 0$. All other sum rules derived in this way could be evaluated due to the form of the sum rule resp. the lack of sufficient experimental numbers only in a very rough way by saturating the integral with one particle contributions (stable particles resp. resonances). The quality of such an approximation depends for a given number of one particle contributions critical on the considered values of t and q_i^2. In the case of arbitrary values for t and q_i^2 it is only possible to saturate sum rules by an infinite number of one-particle contributions. This is connected to the fact that a saturation of a current commutator $[j_{\mu,i}(x), j_{\nu,k}(y)]$ of non-trivial currents by means of a finite number of one-particle intermediate states violates locality.

Two ways out of this difficulty have been proposed:

1. Take the aspect of an infinite number of one-particle contributions serious (Gell-Mann).

2. Extend the contributions of one discrete one-particle intermediate state in a local manner (Schroer, Völkel).

6. Low Energy Theorems

The low energy theorems derived from current algebra belong to two groups:

1. Low energy theorems following from the algebra of charges. As an example let us consider a result on pion-nucleon scattering lengths:

In the derivation of the Adler-Weisberger relations we had to assume that the limit $p_0 \to \infty$ may be interchanged with the q_0-integration resp. that the retarded commutator matrixelement satisfies an unsubtracted dispersion relation. One may ask the question whether it is possible to get a physical relevant result without such an assumption.

Before applying the above mentioned assumption the Adler-Weisberger relation may be put in the following form

$$1 = (2\pi)^2 \frac{\partial}{\partial s} M_R(s, 0)_{s=m^2}$$

with

$$M_R(s, q^2) \equiv 2\pi i \int d^4x\, e^{iqx} \theta(x_0) \langle p | [D_+(x/2), D_-(-x/2)] | p \rangle .$$

Thereby, we have

$$M_R(s, q^2) \propto T_{\pi^- p}(s, q^2)/(q^2 - \mu^2)^2 .$$

Pion-nucleon scattering lengths are defined through the scattering amplitude at threshold

$$a_{\pi^\pm p} \sim T_{\pi^\pm p}\{m + \mu)^2, \mu^2\} = \text{Real}\, T_{\pi^\pm p}\{(m + \mu)^2, \mu^2\} .$$

With PCAC

$$T_{\pi p}(s,\mu^2) \cong T_{\pi p}(s,0)\,,$$

$s \leftrightarrow u$-crossing

$$\mathrm{Real}\,T_{\pi^- p}(s,0) = \mathrm{Real}\,T_{\pi^+ p}(-s+2m^2,0)$$

and analyticity of the scattering amplitude in s we obtain as a first term in a Taylor expansion

$$a_{\pi^- p} - a_{\pi^+ p} \sim \frac{\partial}{\partial s}\,\mathrm{Real}\,T_{\pi^- p}(s,0)|_{s=m^2}\,.$$

By means of the above mentioned form of the Adler-Weisberger relation we finally get

$$a_{\pi^- p} - a_{\pi^+ p} \cong \frac{\mu^4}{2\pi c^2}\,\frac{m\mu}{(m+\mu)} \cong 0.23\,f$$

which has to be compared with the experimental value $(0.252 \pm 0.007)\,f$.

2. Low energy theorems following from ETCR between one charge and one current.

One typical example is the Callan-Treiman relation, i.e. a relation between matrixelements for the decay $K^+ \to \pi_0 + l^+ + \nu$ resp. $K^+ \to l^+ + \nu$.

The hadronic part of the $K^+ \to \pi_0 + l^+ + \nu$ decay matrixelement has the following general form

$$\langle \pi_0(q)|j^{(v)}_{\mu,K^-}|K^+(p)\rangle = (2\pi)^{-3}\,\frac{1}{\sqrt{2}}\,[F^+(t,\mu^2)\,(p+q)_\mu + F^-(t,\mu^2)\,(p-q)_\mu]$$

whereas the hadronic part of the $K^+ \to l^+ + \nu$ decay matrix element looks very simple

$$\langle 0|j^{(A)}_{\mu,K^-}|K^+(p)\rangle = (2\pi)^{-3/2}\,i\,f_K p_\mu\,.$$

By means of the commutator

$$[Q_3^5(x_0), j^{(v)}_{\mu,K^-}(x)] = -\frac{1}{2}\,j^{(A)}_{\mu,K^-}(x)$$

we obtain the relation

$$\frac{f_K}{f_\pi} = F^+(m_K^2,0) + F^-(m_K^2,0)$$

resp. with PCAC for the r.h.s.*

$$\frac{f_K}{f_\pi} = F^+(m_K^2,\mu^2) + F^-(m_K^2,\mu^2)\,.$$

---

* Corrections to this soft pion result have been discussed recently by *B. Stech* (Lectures given at the Niehls Bohr Institute, Copenhagen).

Present experimental numbers are compatible with the Callan-Treiman relation. For a quantitative check more detailed experimental numbers are necessary.

Another example is the $|\Delta\vec{I}|=1/2$ rule for the non-leptonic decays of strange particles:

In the framework of a current-current interaction

$$H^{NL}_{\text{weak}} = \frac{g}{\sqrt{2}}\,\frac{1}{2}(J_\mu J^{\mu^+} + J^+_\mu J^\mu)$$

with

$$J_\mu \equiv :\cos\theta(j^{(v)}_{\mu,1+i2} + j^{(A)}_{\mu,1+i2}) + \sin\theta(j^{(v)}_{\mu,4+i5} + j^{(A)}_{\mu,4+i5})$$

the $|\Delta\vec{I}|=1/2$ rule is not contained as a selection rule because in the product of only charged isospin 1 $(j^{(A,v)}_{\mu,1+i2})$ resp. isospin 1/2 $(j^{(A,v)}_{\mu,4+i5})$ currents an isospin 3/2 part necessarily occurs. We, therefore, conclude that the experimentally found smallness of the $I=3/2$ contribution must be of a dynamical origin. Current algebra leads in particular cases to this $|\Delta\vec{I}|=1/2$ rule in the form of a low energy theorem:

From the current-current interaction and the ETCR postulated above we find immediately

$$[Q, H^{NL}_{\text{weak}}(x)] = [Q^5(x_0), H^{NL}_{\text{weak}}(x)]\,.$$

Therefore, we have, if we consider the example of the $K_{2\pi}$-decay:

$$\langle\pi_i\pi_j|H^{NL}_{\text{weak}}|\rangle \underset{p_\pi\to 0}{\simeq} \langle 0|[Q^5_i[Q^5_j, H^{NL}_{\text{weak}}]]|K\rangle$$

$$= \langle 0|[Q_i[Q_j, H^{NL}_{\text{weak}}]]|K\rangle \sim \langle 0|H^{NL'}_{\text{weak}}|K\rangle$$

i.e.

$$\langle\pi_i\pi_j|H^{NL(3/2)}_{\text{weak}}|K\rangle \underset{p_\pi\to 0}{\to} 0$$

because the K-meson has isospin 1/2.

In summarizing our talk we may state:

Gell-Mann's current algebra has been proved to be an appropriate principle for the formulation of symmetry breaking in strong interactions and/or the universal coupling of hadron currents in weak interactions.

References

1. *Dipole-Sum Rule:*
Blatt, J. M., and *V. F. Weisskopf:* Theoretical nuclear physics, p. 640. New York: John Wiley & Sons 1952.
2. *SU(3)-Symmetry:*
Gell-Mann, M., and *Y. Ne'eman:* The eightfold way. New York-Amsterdam: W. A. Benjamin 1964.

3. *SU(3) × SU(3) Algebra of Currents:*
Gell-Mann, M.: Phys. Rev. **125**, 1067 (1962); — Physics **1**, 63 (1964).
4. *Current Algebra in Perturbation Theory:*
Johnson, K., and *F. E. Low:* Progr. Theor. Phys. Suppl. **37**, **38**, 74 (1966); — *Schroer, B.*, and *P. Stichel:* Commun. Math. Phys. **8**, 327 (1968); — *Stichel, P.:* Lectures given at this school.
5. *Adler-Weisberger Relation:*
Adler, A.: Phys. Rev. Letters **14**, 1051 (1965); — *Weisberger, W. I.:* Phys. Rev. Letters **14**, 1047 (1965); — *Schroer, B.*, and *P. Stichel:* Commun. Math. Phys. **3**, 258 (1966).
6. *Sum Rules à la Fubini:*
Fubini, S.: Nuovo Cimento **43**, 475 (1966).
7. *Saturation of Current Algebra by One-Particle Contributions:*
Gell-Mann, M., D. Horn, and *J. Weyers:* In: Proceedings of the Heidelberg International Conference on Elementary Particles, p. 479. Amsterdam: North-Holland Publ. Co. 1968; — *Völkel, U., B. Schroer*, and *A. H. Völkel:* University of Pittsburgh preprint, February 1968.
8. *Current Algebra and πN-Scattering Lengths:*
Weinberg, S.: Phys. Rev. Letters **17**, 616 (1966).
9. *Callan-Treiman Relation:*
Callan, C. G., and *S. B. Treiman:* Phys. Rev. Letters **16**, 153 (1966).
10. $|\Delta I| = 1/2$*-Rule as Low-Energy Theorem:*
Sugawara, H.: Phys. Rev. Letters **15**, 870 (1965) E 997; — *Suzuki, M.:* Phys. Rev. Letters **15**, 986 (1965).
11. For a detailed list of the original contributions in the field of current algebra kindly refer to the book by *Renner, B.*, which was recently published: Current algebras and their applications. New York: Pergamon Press 1968.
12. *Adler, S. L.*, and *R. F. Dashen:* Current algebra and applications to particle physics. New York-Amsterdam: W. A. Benjamin 1968.

Prof. Dr. *P. Stichel*
Physikalisches Staatsinstitut der Universität Hamburg
2000 Hamburg 50, Luruper Chaussee 149
and Deutsches Elektronensynchrotron (DESY) Hamburg
2000 Hamburg 52, Notkestieg 1

Realisations of a Compact, Connected, Semisimple Lie Group*

JULIUS WESS

Contents

Introduction and Explanations

Before I am going into any details let me first explain all the words I have just used. A *group* is a set G of elements g ($g \in G$) for which a multiplication law is defined:

$$g_1 \cdot g_2 = g \in G$$

such that

1) $(g_1 \cdot g_2)g_3 = g_1(g_2 g_3)$ (associative law);
2) there is an unit element $e: eg = ge = g$;
3) every element g possesses an inverse element g^{-1}:

$$gg^{-1} = g^{-1}g = e\,.$$

If there is a subset H of G ($H \subset G$) which by itself fullfills all these postulates we call this *subset* a *subgroup*. Given a subgroup H we can divide the whole group G into classes, the socalled *cosets* (G/H). A *left coset* being defined as the set of all elements which can be written in the form gh with g fixed and h an arbitrary element of H (symbolically gH).

It is clear that two cosets are either identical or have no element in common. Assume there is one element in two cosets $g_1 h_1 = g_2 h_2$. It follows that $g_1 = g_2 h_2 h_1^{-1} = g_2 h$ and, therefore, $g_1 H = g_2 H$, the two cosets are identical. In each coset we can choose one element ξ to *represent* the whole coset.

The *right cosets* are analogously defined. If now for a particular subgroup H all the left cosets Hg are identical with the right cosets gH, or $gHg^{-1} = H$, we call this subgroup an *invariant subgroup*. A group is

* The lecture is based on some work done together with *S. Coleman* and *B. Zumino*.

called *simple* if it contains no proper invariant subgroup. A group is said to be *semisimple* if none of its proper invariant subgroups are Abelian — a group is called *Abelian* if for any two elements $g_1 g_2 = g_2 g_1$.

We shall specify our groups by the requirement that its elements can be labeled by r continuously varying real parameters $\alpha^1, \ldots, \alpha^r : g = g(\alpha^1, \ldots, \alpha^r)$. Such a group is called a r-parameter *continuous group*. A r-parameter *Lie group* is a r-parameter continuous group with the additional requirement that the parameters of a product γ_i

$$g(\alpha_1, \ldots, \alpha_r)\, g(\beta_1, \ldots, \beta_r) = g(\gamma_1, \ldots, \gamma_r)$$
$$\gamma_i = \gamma_i(\alpha_1, \ldots, \alpha_r, \beta_1, \ldots, \beta_r)$$

are analytic functions of the parameters $\alpha_1, \ldots, \alpha_r, \beta_1, \ldots, \beta_r$.

Finally, if the parameters of a Lie group vary over a finite range, the group is said to be *compact*, if this range is connected, the group is called *connected*.

We shall talk about a *realisation* of the group if we deal with an n-dimensional real analytic manifold on which the group acts as group of transformations. We assume that these transformations are analytic in the parameters of the group and manifold. These topological restrictions are entirely justified by the problems we have in mind [1] — to construct invariant Lagrangians as functions of fields which span the manifold. These functions are always treated in power series expansion and it is sufficient to consider invariance under infinitesimal transformations.

With x we denote the points of the manifold M and we write $x' = gx$ if the group element g maps the point x into the point x'. We ask for all group elements S which leave a given, fixed point x_0 invariant, i.e. $x_0 = s x_0$. Clearly, these elements form a group S; the socalled *stability group* of the point x_0. Let now x be a point which can be connected with the point x_0 by a group element $g : x = g x_0$. It is obvious that all group elements which belong to the same *left coset* of the group G with respect to the group S connect the two points $x, x_0 : x = g s x_0 = g x_0$. Moreover, if two elements g_1 and g_2 connect the two points x, x_0, then they belong to the same left coset: $x = g_1 x_0 = g_2 x_0$, $g_1^{-1} g_2 x_0 = x_0$ or $g_1^{-1} g_2 = h$, $g_2 = g_1 h$.

Thus we have a one to one correspondence between the left cosets G/S and the points x of M which can be connected with x_0. This submanifold of M which we call a *layer* can, therefore, be labeled with the parameters of the group element ξ which represents the particular coset. We can now decompose our whole manifold into such layers. Each layer by itself is minimal in the sence that it is *transitive*, any two points can be connected by a group element $x_1 = g_1 x_0$, $x_2 = g_2 x_0$, $x_2 = g_2 g_1^{-1} x_1$.

Thus our manifold M is isomorphic to a certain composition of layers, each of which is again isomorphic to a certain coset of the group G.

Our problem, to find all possible realisations of the group G can in this sence be reduced to the problem of finding all subgroups of the group G and labeling the respective cosets.

Unfortunately, we have to complicate our problem at this point by an additional requirement on our realisations. We demand that a certain subgroup H of G can be *represented* linearly on the manifold M.

If this subgroup is the group G itself we are confronted with the problem of finding all *linear representations* of a compact, connected, semisimple Lie group, a problem we are all familiar with and I have only to quote a few results. All the finite dimensional representations are known and they are equivalent to unitary representations. We denote the matrices of our irreducible unitary representation with $D^{(\varrho)}_{\nu\mu}(g)$; ϱ is a discrete parameter which labels the respective representation.

The following question arises: given a realisation of the group G, what are the conditions that the realisation can be linearized. The answer is given by the following criterion:

If there is a point which is left invariant under the transformations: $gx_0 = x_0$, then it is possible to introduce coordinates in the neighborhood of this point which transform linearly. We call this point the origin.

Obviously, for any representation such a point — the origin exists.

We now assume that for an arbitrary realisation we have found such a point x_0 in the manifold M. We can find a neighborhood of the point x_0 which is left invariant under the action of G. This is the case because G is a compact, connected Lie group. Within this neighborhood we introduce coordinates which we again denote by x and we take x_0 as the origin $x_0 \equiv 0$. The actions of the group on x can now be expanded in a power series

$$g \cdot x = D(g)\, x + o(x^2)\,.$$

First we observe that $D(g)$ cannot be identically zero. This follows from continuity and from the fact that $ex = x$. Moreover, $D(g)$ is a linear representation of G.

$$\begin{aligned} g_1 g_2 \cdot x &= D(g_1 g_2)\, x + o(x^2) \\ &= g_1 \{D(g_2)\, x + o(x^2)\} = D(g_1)\,\{D(g_2)\, x + o(x^2)\} + o(x^2) \\ &= D(g_1)\, D(g_2)\, x + o(x^2)\,. \end{aligned}$$

Now we introduce coordinates y as follows

$$y = \int_G \mathrm{d}\mu(g)\, D^{-1}(g)\, g \cdot x\,. \tag{1}$$

The measure $\mathrm{d}\mu(g)$ is an abbreviation for $\mathrm{d}\mu(g) = \sigma(\alpha_1, \ldots, \alpha_r)\, \mathrm{d}\alpha_1, \ldots, \mathrm{d}\alpha_r$. The function $\sigma(\alpha_1, \ldots, \alpha_r)$ can for compact groups always be chosen to be left and right invariant

$$d\mu(g_0 g) = d\mu(g g_0) = d\mu(g)$$

and to be normalized:

$$\int_G d\mu(g) = 1 .$$

Thus $y = x + o(x^2)$, the Jacobian determinant $|\delta y/\delta x|$ is equal to one at the origin and the transition from x to y is an allowed change of coordinates. If we now study the action of G on the coordinates y we find

$$\begin{aligned} g_0 \cdot y &= \int_G d\mu(g) D^{-1}(g) g(g_0 x) \\ &= \int_G d\mu(g g_0) D(g_0) D^{-1}(g_0) D^{-1}(g) g g_0 x \\ &= D(g_0) \int_G d\mu(g) D^{-1}(g) g x = D(g_0) y . \end{aligned}$$

This proves the criterion and gives an explicit prescription how to find the new coordinates y that transform linearly.

We are now prepared to deal with the general case, i.e. that a certain subgroup H of G should be represented linearly on M (for simplicity we assume that H is continuous). A necessary and sufficient condition is that there exists a point on M invariant under H. We take this point as origin. The stability group of the origin is the group H. The layer which we obtain by the action of G on the origin is isomorphic to the coset G/H and the parameter of the group elements representing the cosets can be used as coordinates on this layer. We choose a parametrisation of the cosets which is continuous and has the property that the parameters transform linearly under H. That this is always possible is best demonstrated by an explicit construction.

We denote the generators of H with V_i $(i = 1, \ldots, r-d)$ and the remaining generators of G with A_e $(e = 1, \ldots, d)$. The generators should be chosen such that they form a complete set of generators of G, orthonormal with respect to the *Cartan inner product*. This means the following: Given the structure constants of a Lie group $[x_\varrho x_\sigma] = c^k_{\varrho\sigma} x_k$ we can form the metric tensor $g_{\mu\nu} = c^\beta_{\mu\alpha} c^\alpha_{\nu\beta}$ and we can define the scalar product $(x, y) = x_\mu g_{\mu\nu} y_\nu$. Since $g_{\mu\nu}$ is a real symmetric matrix it can be brought into diagonal form by a suitable linear transformation of the generators

$$g_{\mu\nu} = \varepsilon_\mu \delta_{\mu\nu} , \qquad \varepsilon_\mu = \pm 1 .$$

The necessary and sufficient condition for a Lie algebra to be semisimple is that $\det g \neq 0$. The necessary and sufficient conditions for a semisimple Lie algebra to be compact is that the matrix $g_{\mu\nu}$ be negative definite. Thus we can choose the generators such that $g_{\mu\nu} = -\delta_{\mu\nu}$.

Group elements in the neighborhood of the identity in G — and for Lie groups it suffices to study such elements — can be decomposed uniquely into a product of the form

$$g = e^{\sum_1^d \xi_i A_i} e^{\sum_1^r u_e V_e} .$$

This follows from the fact that

$$e^{\xi_i A_i} e^{u_e V_e} = e^{\xi_i A_i + u_e V_e + \frac{1}{2}[\xi_i A_i, u_e V_e] + [\text{higher commutators}]}$$
$$= e^{\bar{\xi}_i(\xi, u) A_i + \bar{u}_e(\xi, u) V_e} .$$

The Jacobian determinant of the transformation

$$\begin{cases} \bar{\xi}_i = \bar{\xi}_i(\xi, u) \\ \bar{u}_e = \bar{u}_e(\xi, u) \end{cases}$$

is equal to one at the origin, therefore, the transformation can be inverted in a neighborhood of the origin

$$\xi_i = \xi_i(\bar{\xi}, \bar{u}) , \qquad u_e = u_e(\bar{\xi}, \bar{u}) ,$$

which proves the uniqueness of the decomposition.

We take the variables ξ as coordinates on our manifold and study their transformation properties under the group.

For every element $g_0 \in G$ we can write

$$g_0 e^{\xi_i A_i} = e^{\xi'_i A'_i} e^{u'_e V_e} . \tag{2}$$

The parameters ξ' and u' are functions of ξ and g.

If g_0 belongs to H, $g_0 = h_0$ we can write

$$h_0 e^{\xi_i A_i} = h_0 e^{\xi_i A_i} h_0^{-1} h_0 . \tag{3}$$

The transformation

$$g_0 e^{\xi_i A_i + u_e V_e} g_0^{-1} = e^{\xi'_i A_i + u'_e V_e}$$

defines the *adjoint* representation of the group G. The Cartan scalar product is left invariant under these transformations. If we restrict the transformations to H, $g_0 = h_0 \in H$; the subspace spanned by the vectors $\Sigma u_e V_e$ is invariant by itself. Consequently, the orthogonal subspace, spanned by the vector $\Sigma \xi_i A_i$ is also invariant and we can write

$$h_0 e^{\xi_i A_i} h_0^{-1} = e^{\xi'_i A_i}$$

with

$$\xi'_i = D_{ij}^{(b)}(h_0) \xi_j .$$

$D^{(b)}$ is a linear representation of H, contained in the adjoint representation of G. We can rewrite formula (3)

$$h_0 e^{\xi_i A_i} = e^{\xi'_i A_i} e^{u'_e V_e} , \tag{4}$$

where we have made the identification $h_0 \equiv e^{u'_e V_e}$. In this special case u' is independent of ξ. This proves that our coordinates ξ_i have the desired properties — to be continuous and to transform linear under H.

Using the above formulas we can find a whole class of realisations of G, which, when restricted to H, become representations. Let $D(h)$ be an arbitrary representation of H. Then

$$g_0: \xi \to \xi', \qquad \psi \to D(e^{u'_e V_e})\psi \tag{5}$$

is a realisation which has the desired properties. Eq. (2) defines the parameters ξ' and u'. That (5) defines a realisation can be easily checked:

$$g_0 e^{\xi A} = e^{\xi' A} e^{u' V},$$
$$g_1 e^{\xi' A} = e^{\xi'' A} e^{u'' V},$$
$$g_1 g_0 e^{\xi A} = e^{\xi'' A} e^{u'' V} e^{u' V} = e^{\xi''' A} e^{u''' V}.$$

Thus: $e^{u''' V} = e^{u'' V} e^{u' V}$ and since D is a representation

$$D(e^{u''' V}) = D(e^{u'' V}) D(e^{u' V}).$$

Observe that u' will in general be a function of ξ. Therefore, it is only possible to define the transformation (5) on the manifold (ξ, ψ). That the realisation becomes linear under H follows immediately from (4).

$$h_0: \quad \xi' = D^b(h)\xi, \qquad \psi' = D(h)\psi.$$

Next we shall show that manifolds on which a realisation of g is defined that becomes linear under H can always be parametrized in such a way that in the neighborhood of the origin the group transformations can be written in the form (5). It is, therefore, justified to call (5) a standard form for such realisations. If $D(h)$ is reducible we shall write our standard forms such that $D(h)$ is fully reduced. Our vectors ψ can then be decomposed into vectors with a smaller number of components ($\psi = \Sigma \oplus \psi^a$) and these vectors do not mix under the group transformations.

To prove our statement we remember that we have defined an origin of our n-dimensional real analytic manifold and that we have introduced the group parameters $\xi_1, \ldots, \xi_d$ as coordinates on the layer which containes the origin. We introduce $n-d$ other coordinates on M and write them as a real vector ψ. A point of M (in the neighborhood of the origin) has the pair (ξ, ψ) as coordinates. We study now the action of the group From the very definition of the parameters ξ follows that

$$g_0(\xi, 0) = (\xi'(\xi, g_0), 0),$$

where ξ' is given by Eq. (2). By assumption, the coordinates can be chosen such that the subgroup H acts linearly. Real representations of the compact group H can always be brought into orthogonal form by a suitable choice of coordinates. Given an invariant subspace (the layer) it is, therefore, always possible to choose the complementary space also invariant. This means that we can choose the coordinate ψ always in such a way that for $h_0 \in H$:

$$h_0(\xi, \psi) = (D^b(h_0)\xi\,,\; D(h_0)\psi)\,.$$

Finally, we introduce new coordinates $(\xi, \psi)^*$ by the following definition:

$$(\xi, \psi)^* = e^{\xi_e A^e}(0, \psi)\,.$$

This is an allowable change of coordinates in some neighborhood of the origin, since (6) defines an analytic mapping of the coordinates whose Jacobian determinant does not vanish at the origin. As a special case Eq. (6) contains

$$(0, \psi)^* = (0, \psi) \quad \text{and} \quad (\xi, 0)^* = (\xi, 0)\,.$$

The action of the group G on this new set of coordinates is the following

$$\begin{aligned} g(\xi, \psi)^* &= g\, e^{\xi A}(0, \psi)^* = e^{\xi_i A_i} e^{u' V}(0, \psi)^* \\ &= e^{\xi A}(0, D(e^{u' V})\psi)^* = (\xi', D(e^{u' V})\psi)^*\,. \end{aligned}$$

The parameters u' and ξ' are given by Eq. (2). To compare this transformation law with our standard form (5) we only have to remember that a real orthogonal representation is either truly real or that it can be brought into a unitary form by combining real vectors into complex vectors.

Thus we have proven our previous statement and solved the problem to find all realisations of a compact connect semi-simple Lie group which can be linearized when restricted to a continuous subgroup H.

In addition we have developed a method which allows us to transform such realisations into a standard form. In order to give explicit formulas it is necessary to solve the Eq. (2) for ξ' and u'. This can be simplified in the case in which the group G admits the automorphism $R: g \to R(g)$ such that $V_i \to V_i$, $A_e \to -A_e$. We apply the automorphism to the relation (2)

$$g_0 e^{\xi_i A_i} = e^{\xi'_i A_i} e^{u_e V_e} \to R(g_0) e^{-\xi_i A_i} = e^{-\xi'_i A_i} e^{u'_e V_e}\,.$$

We can now eliminate u' and obtain

$$g_0 e^{2\xi_i A_i} R(g_0^{-1}) = e^{2\xi' A}\,.$$

In this formula it becomes obvious that $\xi \to \xi'$ is a realisation of the group which becomes linear when restricted to the subgroup H.

$$\begin{aligned} g'_0 g_0 e^{2\xi A} R((g'_0 g_0)^{-1}) &= g'_0 g_0 e^{2\xi A} R(g_0^{-1}) R(g_0'^{-1}) \\ &= g'_0 e^{2\xi' A} R(g_0'^{-1}) \end{aligned}$$

and $h e^{2\xi A} h^{-1} = e^{2\xi' A}$ defines the adjoint representation.

Relations between Linear and Nonlinear Transformations

As an example for a manifold on which H can be represented linearly let us consider the manifold (ξ, Ξ_α); ξ transforms in the usual way, Ξ_α should transform according to a linear irreducible and unitary representation $D(g)$ of G:

$$\Xi'_\alpha = D_{\alpha\beta}(g)\,\Xi_\beta\,.$$

We want to find a parametrisation of this manifold which corresponds to our standard form. All we have to do for this purpose is to introduce the new coordinates according to (6):

$$(\xi, \psi)^* = e^{\xi A}(0, \psi)^* = e^{\xi A}(0, \psi) = (\xi, \Xi)\,.$$

This relation means that the point (ξ, Ξ) characterized by its "old" coordinates, gets the "new" coordinates (ξ, ψ) where ψ is the old coordinate of that point into which the point (ξ, Ξ) moves under the group transformation $e^{-\xi A}$ (this is the transformation which transforms $\xi \to 0$).

Let us now verify explicitely that the coordinates

$$(\xi, \psi_\alpha),\ \psi_\alpha = D_{\alpha\beta}(e^{-\xi A})\,\Xi_\beta$$

transform like our standard form:

$$\begin{aligned}\psi'_\alpha &= D_{\alpha\beta}(e^{-\xi' A})\,\Xi'_\beta = D_{\alpha\beta}(e^{-\xi' A} g)\,\Xi_\beta \\ &= D_{\alpha\beta}(e^{-\xi' A} g\, e^{\xi A})\,\psi_\beta = D_{\alpha\beta}(e^{u' V})\,\psi_\beta\,.\end{aligned}$$

The last relation follows from (2). In case that the representation $D(h)$ of the group H is reducible the vectors ψ_α break up into a set of vectors $\psi_\alpha^{(a)}$.

$$(\xi, \psi) \sim \left(\xi, \sum_a \oplus\, \psi^{(a)}\right)\,.$$

The group then provides no connection between the different representations of H. The dimensionality of the multiplets is that of the irreducible representations of H.

It is clear that the above construction can be inverted. If we have a standard form $\left(\xi, \sum_a \oplus\, \psi^a\right)$ and if in the direct sum $\sum_a \oplus\, \psi^a$ there occur exactly those representations of H which occur in the representation $D(g)$ of G we can introduce the vector

$$\Xi_\alpha = D_{\alpha\beta}(e^{\xi A})\,\psi_\beta \tag{7}$$

and show that

$$\Xi'_\alpha = D_{\alpha\beta}(g)\,\Xi_\beta\,.$$

Here we have not changed the dimensionality of the manifold and the transformation (7) can be inverted. From our definitions of equivalence for realisations follows that representations of G which have the same "content" of representations of H are equivalent if also the parameters ξ are part of the manifold. From the point of view of a Lagrangian theory this means that there is no way to distinguish such "equivalent" representations as long as a field ξ is present.

A slightly more general problem is to find those functions f_α of ξ and ψ which transform according to linear representations of the group G.

$$g: f_\alpha(\xi, \psi) \to f_\alpha(\xi', \psi') = \sum_\beta D_{\alpha\beta}(g_0) f_\beta(\xi, \psi).$$

We observe that it is sufficient to study the case in which f_α is a linear function of ψ

$$f_\alpha(\xi, \psi) = \sum F_{\alpha r}(\xi) \psi_r$$

since the general case can be reduced to this only by first solving the Clebsch-Gordan problem for the subgroup H and the tensor products of the ψ's. We write, therefore:

$$\sum_r F_{\alpha r}(\xi') \psi'_r = \sum_{\beta, s} D_{\alpha\beta}(g_0) F_{\beta s}(\xi) \psi_s.$$

We show first as a necessary condition: Only such representations of G can be realized on the space of functions $\sum_r F_{\alpha r}(\xi) \psi_r$ which contain the representation $R(h)$ of H under which ψ transforms at least once. If we put $\xi = 0$ in the above equation and restrict the transformation to the subgroup H we find

$$\sum_{rs} F_{\alpha r}(0) R_{rs}(h) \psi_s = \sum_{\beta s} D_{\alpha\beta}(h) F_{\beta s}(0) \psi_s$$

because $\xi = 0$ is left invariant by H. It follows from Schur's Lemma that the matrix $F_{\alpha r}(0)$ is either identical zero or that $D(h)$ contains the irreducible representation $R(h)$. The first alternative is impossible because $F_{\alpha r}(0) = 0$ implies $F_{\alpha r}(\xi) = 0$ for all ξ, this follows immediately from the transitivity of the submanifold labeld by ξ.

The condition just mentioned is also sufficient. To show this we study the decomposition $g = e^{\xi A} e^{uV}$ of the group element g in the representation $D(g)$. A suitable basis can be found such that

$$D(g) = D(e^{\xi A}) \begin{pmatrix} R(e^{uV}) & 0 \\ 0 & \tilde{D}(e^{uV}) \end{pmatrix}.$$

Now:

$$\begin{aligned} D(g_0) D(g) &= D(g_0 e^{\xi A} e^{uV}) = D(e^{\xi' A} e^{u' V} e^{uV}) \\ &= D(e^{\xi' A}) \begin{pmatrix} R(e^{u' V} e^{uV}) & 0 \\ 0 & \tilde{D}(e^{u' V} e^{uV}) \end{pmatrix} \end{aligned}$$

or

$$\sum_{\beta} D_{\alpha\beta}(g_0) D_{\beta s}(e^{\xi A}) = \sum_{r} D_{\alpha r}(e^{\xi' A}) R_{rs}(e^{u' V})$$

if the indices s and r run only through the labels of the first subspace (in our basis) invariant under H.

The functions $\sum_{s} D_{\alpha s}(e^{\xi A})\psi_s$ have the desired property:

$$\begin{aligned}\sum_{s} D_{\alpha s}(e^{\xi' A})\psi'_s &= \sum_{sr} D_{\alpha s}(e^{\xi' A}) R_{sr}(e^{u' V})\psi_s \\ &= \sum_{\beta r} D_{\alpha\beta}(g_0) D_{\beta r}(e^{\xi A})\psi_r .\end{aligned}$$

The same argument shows that if the representation D reduces and contains the representation $R(h)$ m times when restricted to the subgroup, then one can construct m sets of linearly independent functions $F_{\alpha r}(\xi)$.

The problem to construct such functions arises e.g. if we want to add a term to the lagrangian which breaks the symmetry in such a way that it transforms according to a component of a linear representation. Such a term has then to be constructed as a function of the fields. It is then of special interest to construct such functions of the variables ξ alone. This is a particular case of the problem just studied — let the representation $R(h)$ be the identity representation and replace ψ by a constant. It is also easy to see that, if there is another subgroup H_1, of G, containing H and if the representation D is reducible and contains the identity also when it is taken on H_1, then the functions $D_{\alpha 1}(e^{\xi A})$ do not really depend on all ξ_e, but only on those which are not associated with the generators A_e which belong to the larger subgroup H_1. Thus, in particular, if D is the identical representation the function will not depend on any of the fields ξ, it will be a constant.

We ask as a final problem: given two linear representations of G, R^1 and R^2, $\Xi^{(1)'}_\alpha = R^{(1)}_{\alpha\beta}(g)\Xi^{(1)}_\beta$, $\Xi^{(2)'}_\alpha = R^{(2)}_{\alpha\beta}\,\Xi^{(2)}_\beta$ under what circumstances is there a function $F_{\mu\beta}(\xi)$ such that $F_{\mu\beta}(\xi)\Xi^{(1)}_\beta$ transforms as $\Xi^{(2)}$?

We show that it is necessary and sufficient that $\overline{R^{(1)}} \times R^{(2)}$ contains in its reduction one of the representations which can be realized on functions of ξ alone. Necessary: by assumption

$$R^{(2)}_{\mu\nu}(g) F_{\nu\beta}(\xi)\Xi_\beta = F_{\mu\alpha}(\xi') R^1_{\alpha\beta}(g)\Xi_\beta$$

or

$$\begin{aligned}F_{\mu\alpha}(\xi') &= R^2_{\mu\nu}(g) F_{\nu\beta}(\xi)\, |R^{(1)}(g)|^{-1}_{\beta\alpha} \\ &= R^{(2)}_{\mu\nu}(g)\, \overline{R^{(1)}_{\alpha\beta}}(g)\, F_{\nu\beta}(\xi) .\end{aligned}$$

Sufficient: We define the Clebsch-Gordon coefficients $\langle \alpha 1; \nu 2|\sigma r\rangle$ for the reduction:

$$\langle \alpha\bar{1}|\langle \nu, 2| = \sum_{\sigma r} \langle \alpha, \bar{1};\, \nu, 2|\sigma r\rangle \langle \sigma, r| .$$

By assumption, for at least one value r_0 of r there exists a-set of functions $f_\sigma(\xi)$ such that

$$f_\sigma(\xi') = R^{(r_0)}_{\sigma\varrho}(g) f_\varrho(\xi).$$

From the completeness of the Clebsch-Gordon coefficients follows that

$$\overline{R^{(1)}_{\alpha\beta}} R^{(2)}_{\mu\nu} \langle \beta, \bar{1}; \nu, 2 | g, 1 \rangle = \langle \alpha \bar{1}; \mu 2 | \sigma, r \rangle R^{(r)}_{\sigma\beta}.$$

We define now

$$F_{\mu\alpha}(\xi) = \langle \alpha, 1, \mu, 2 | \sigma r_0 \rangle f_\sigma(\xi)$$

and we see that this function has the desired properties:

$$\begin{aligned} F_{\mu\alpha}(\xi') &= \langle \alpha, \bar{1}; \mu 2 | \sigma, r_0 \rangle R^{(r_0)}_{\sigma\varrho} f_\varrho(\xi) \\ &= R^{(1)}_{\alpha\beta} R^{(2)}_{\mu\nu} \langle \beta, 1; \nu, 2 | \varrho, r_0 \rangle f_\varrho(\xi) \\ &= \overline{R^{(1)}_{\alpha\beta}} R^{(2)}_{\mu\nu} F_{\nu\beta}(\xi) = R^{(2)}_{\mu\nu} F_{\nu\beta}(\xi) \, [R^{(1)}]^{-1}_{\beta\alpha}. \end{aligned}$$

References

1. *Cronin, J.:* Phys. Rev. **161**, 1483 (1967). — *Weinberg, S.:* Phys. Rev. Letters **18**, 188 (1967). — *Schwinger, J.:* Phys. Letters **24** B, 473 (1967). — *Wess, J.*, and *B. Zumino:* Phys. Rev. **163**, 1727 (1967). — *Zumino, B.:* Phys. Letters **25** B, 349 (1967). — *Lee, B.*, and *H. T. Nieh:* Phys. Rev. **166**, 1507 (1968). — *Coleman, S., J. Wess*, and *B. Zumino:* The Structure of Phenomenological Lagrangians I, to be published in Phys. Rev.

Prof. Dr. *Julius Wess*
Institut für Theoretische Physik
der Universität
7500 Karlsruhe, Kaiserstraße 12

Problems in Vector Meson Theories

W. ZIMMERMANN*

Contents

In these lectures we shall be concerned with a number of difficulties which occur in any Lagrangian field theory involving vector mesons. Unfortunately it has not yet been possible to resolve these problems for Yang-Mills fields or charged vector mesons in general. Therefore, we will restrict ourselves to the case of neutral vector mesons in which case one can give a fairly complete analysis of the problems. In Part I we will discuss the model of a massive vector field coupled to a conserved current. Part II concerns models of two vector fields which are coupled to the same current.

I. One Massive Vector Field Coupled to a Conserved Current

1. The Proca-Wentzel Formulation

Let us consider a neutral, massive vector field V^μ which is coupled to the conserved current of a spinor field φ. In the Proca-Wentzel

* On leave of absence from Courant Institute of Mathematical Sciences, New York University, New York, N.Y.

formulation the Lagrangian is [1]

$$\mathscr{L} = -\frac{1}{4} V_{u\mu\nu} V_u^{\mu\nu} + \frac{1}{2} m_0^2 V_{u\mu} V_u^\mu - g_0 j^\mu V_{u\mu} + \mathscr{L}_\varphi , \tag{1}$$

$$j^\mu = \bar{\varphi}\gamma^\mu \varphi .$$

$\mathscr{L}_\varphi$ denotes the Lagrangian of the free Dirac field. All fields are unrenormalized what is indicated by the subscript u. Renormalized fields are introduced in the usual way by

$$V^\mu = Z_V^{-\frac{1}{2}} V_u^\mu , \quad \varphi = Z_2^{-\frac{1}{2}} \varphi_u , \quad g = Z_V^{\frac{1}{2}} g_0 . \tag{2}$$

The field equations following from this Lagrangian are

$$-\partial_\nu V^{\mu\nu} + m_0^2 V^\mu = Z_V^{-1} g j^\mu , \tag{3}$$

$$(i\gamma^\mu \partial_\mu - M_0)\varphi = g\gamma^\mu V_\mu \varphi . \tag{4}$$

The current is conserved

$$\partial_\mu j^\mu = 0 . \tag{5}$$

The main advantage of the Proca-Wentzel formulation is that it is not necessary to impose a subsidiary condition. For if $m_0^2 \neq 0$ the relation

$$\partial_\mu V^\mu = 0 \tag{6}$$

follows by taking the divergence of (3) and using current conservation (5). Hence the subsidiary condition is fulfilled automatically.

The system is quantized by applying the usual rules. The pairs of canonically conjugate variables are

$$(V_u^k, V_u^{0k}) , \quad (\varphi_u, \varphi_u^*) .$$

This leads to the equal time commutators

$$[V^k(x), V^{0l}(y)] = i Z_V^{-1} \delta_{kl} \delta_3(x-y) , \tag{7}$$

$$\{\varphi_\alpha(x)\, \varphi_\beta^*(y)\} = Z_2^{-1} \delta_{\alpha\beta} \delta_3(x-y) . \tag{8}$$

The remaining commutators between canonical variables vanish. The time component V_u^0 cannot be used as a canonical variable since

$$\frac{\partial \mathscr{L}}{\partial \dot{V}^0} = 0 .$$

The canonical commutator (7) of the meson field implies a sum rule for Z_V^{-1}. The general form of the 2-point function

$$\langle V_\mu(x) V_\nu(y) \rangle_0 = (\Box g_{\mu\nu} - \partial_\mu \partial_\nu) \int d\kappa^2 \frac{\sigma(\kappa^2)}{\kappa^2} \Delta_{\kappa^2}(x-y) \tag{9}$$

yields

$$i \langle V^k(x) V^{0k}(y) \rangle_0 = \int d\kappa^2 \sigma(\kappa^2) \delta_3(x-y) , \quad x^0 = y^0 .$$

By comparison with (7) we obtain the sum rule

$$Z_V^{-1} = \int d\kappa^2 \sigma(\kappa^2) = 1 + \int_{\kappa_0^2}^{\infty} d\kappa^2 \varrho(\kappa^2), \qquad \sigma = \delta(\kappa^2 - m^2) + \varrho. \tag{10}$$

There is another important sum rule which is due to *Johnson* [2]. It states that

$$\frac{1}{m_0^2 Z_V} = \frac{1}{m^2} + \int_{\kappa_0^2} \frac{d\kappa^2}{\kappa^2} \varrho(\kappa^2). \tag{11}$$

In order to derive this sum rule one apparently needs some information on the commutator $[V^k, V^0]$. For the vacuum expectation value

$$i\langle[V^k(x) V^0(y)]\rangle_0 = \int d\kappa^2 \frac{\sigma(\kappa^2)}{\kappa^2} \delta_3(x-y)$$

represents the right hand side of Johnson's sum rule. In his original derivation Johnson determined the commutator $[V^k, V^0]$ from the canonical commutation relations. To this end one solves the field equation for V^0

$$V^0 = \frac{1}{m_0^2}(g j^0 + \partial_k V^{k0}),$$

which yields

$$[V^k V^0] = \frac{1}{m_0^2}[V^k j^0] + \frac{1}{m_0^2}[V^k, \partial_l V^{l0}].$$

Since

$$j^0 = \bar{\varphi}\gamma^0\varphi \quad \text{and} \quad [V^k, \varphi] = 0$$

it seems obvious that $[j^0, V^k] = 0$. With this we have

$$[V^k(x) V^0(y)] = \frac{1}{m_0^2 Z_V} \partial^k \delta_3(x-y), \qquad x^0 = y^0 \tag{12}$$

implying the sum rule (11).

At this point we meet the first difficulty. Johnson already pointed out that the relation

$$[j^0(x), V^k(y)] = 0, \qquad x^0 = y^0 \tag{13}$$

is by no means obvious and does not follow from the canonical commutation relations. First of all the operator product $\bar{\varphi}\gamma^0\varphi$ is a very singular object and one may doubt whether it is possible to evaluate the commutator by commuting V^μ with φ and $\bar{\varphi}$ separately. But one can actually prove that this argument fails even if the theory did not contain any divergencies [2]. For, if

$$[j^0(x), V^k(y)] = 0\,, \qquad x^0 = y^0\,, \tag{13}$$

is believed on the basis of the canonical rules one has also to accept the relation

$$[j^l(x), V^{0k}(y)] = 0\,, \qquad x^0 = y^0\,, \tag{14}$$

since the canonical variable V^{0k} commutes with φ and $\bar{\varphi}$. It is easy to see that (13) and (14) lead to a contradiction. We begin by determining the commutator

$$[\dot{V}^k, j^l] = [V^{0k}, j^l] - [\partial^k V^0, j^l]\,.$$

Using (14) and

$$\langle[\partial^k V^0, j^l]\rangle_0 = \langle[\partial^k j^0, V^l]\rangle_0 = 0$$

we obtain

$$\langle[\dot{V}^k, j^l]\rangle_0 = 0\,. \tag{15}$$

Comparing (15) with the general form (9) of the two-point function we find the sume rule

$$Z_V^{-1} m_0^2 = \int d\kappa^2\, \kappa^2\, \sigma(\kappa^2)\,. \tag{16}$$

The three sum rules (10), (11), and (16) combined yield

$$\int \frac{(m_0^2 - \kappa^2)}{\kappa^2}\, \sigma(\kappa^2)\, d\kappa^2 = 0\,,$$

which implies

$$\sigma = \delta(\kappa^2 - m_0^2)\,.$$

Hence the result is that (13) and (14) together can only be fulfilled for a free field. One concludes that

$$[j^0 V^l] = 0$$

is an additional assumption which goes beyond the ordinary quantization rules.

It is clearly a disadvantage of the Proca-Wentzel formulation that the canonical commutation relations are not sufficient for deriving the Johnson sum rule. Later on we will see that (13) is not even correct if the bare meson mass is infinite.

2. Indefinite Metric Formulation

The difficulties which we found in the preceding section can be resolved if we use a different formulation of the same model which is in closer analogy to electrodynamics. The Lagrangian of this formulation is [3—6]

$$\tilde{\mathscr{L}} = \mathscr{L}_C + \mathscr{L}_\psi - g_0 \hat{j}_\mu C_u^\mu ,$$

$$\mathscr{L}_C = -\frac{1}{4} C_{u\mu\nu} C_u^{\mu\nu} + \frac{1}{2} m_0^2 C_{u\mu} C_u^\mu - \frac{1}{2} \frac{m_0^2}{m^2} (\partial_\mu C_u^\mu)^2 , \tag{17}$$

$$C_{u\mu\nu} = \partial_\mu C_{u\nu} - \partial_\nu C_{u\mu} , \qquad \hat{j}_\mu = \overline{\psi}_u \gamma_\mu \psi_u .$$

$\mathscr{L}_\psi$ denotes the Lagrangian of a free Dirac field ψ_u. The field equations are

$$(i\gamma^\mu \partial_\mu - M_0)\psi = g\gamma^\mu C_\mu \psi , \tag{18}$$

$$-\partial_\nu C^{\mu\nu} + m_0^2 C^\mu + \frac{m_0^2}{m^2} \partial^\mu \partial_\nu C^\nu = g Z_V^{-1} \hat{j}_\mu , \tag{19}$$

$$\hat{j}^\mu = Z_2' \overline{\psi} \gamma^\mu \psi , \qquad \partial_\mu \hat{j}^\mu = 0 \tag{20}$$

for the renormalized fields

$$C^\mu = Z_V^{-\frac{1}{2}} C_u^\mu , \qquad \psi = Z_2'^{-\frac{1}{2}} \psi_u , \qquad g = Z_V^{\frac{1}{2}} g_0 . \tag{21}$$

Current conservation implies

$$(\Box + m^2) C = 0 \tag{22}$$

for the divergence

$$C = \frac{1}{m} \partial_\mu C^\mu . \tag{23}$$

Hence the divergence C is a free field. Since the renormalized operator C is supposed to have finite matrix elements the parameter m must be finite and we may choose it as the renormalized mass. The canonical commutation relations of the system imply

$$[C(x), C^\mu(y)] = -\frac{m}{Z_V m_0^2} i\partial^\mu \Delta(x-y) , \tag{24}$$

$$[C(x), C(y)] = -\frac{m^2}{Z_V m_0^2} i\Delta(x-y) . \tag{25}$$

The commutator of the C-field has the wrong sign, therefore one must introduce an indefinite metric in order to have a positive definite energy for the particles associated with the C-field. The space $\mathscr{H}_{\text{phys}}$ of physical state vectors is defined by

$$C^{+}(x)\Phi = 0$$

with $^{+}$ denoting the positive frequency part.

3. Gauge Invariance

It is easy to see that the field equations are invariant under coordinate dependent gauge transformations, despite of the mass term. The gauge transformation

$$\begin{aligned} &C_\mu \to C_\mu + \frac{1}{m}\partial_\mu \Lambda\,, \\ &\psi \to e^{-i\frac{g}{m}\Lambda}\psi \end{aligned} \tag{26}$$

leaves the field equations invariant provided Λ is a solution of the Klein-Gordon equation for the mass m.

The fields of the Proca-Wentzel formulation can now be defined by a special operator gauge transformation. We set

$$\Lambda = C = \frac{1}{m}\partial_\mu C^\mu$$

and define [8]

$$V^\mu = C^\mu + \frac{1}{m}\partial^\mu C\,, \tag{27}$$

$$\varphi = c\,e^{-i\frac{g}{m}C}\psi\,, \tag{28}$$

where c is a suitable normalization constant. V^μ and φ satisfy the field equations of the Proca-Wentzel formulation. Moreover, [7]

$$[V^\mu(x), C(y)] = 0\,, \qquad [\varphi(x), C(y)] = 0 \tag{29}$$

at all times which shows that $V^\mu(x)$, $\varphi(x)$ leave the physical subspace invariant. Hence we may restrict $V^\mu(x)$, $\varphi(x)$ to the subspace $\mathscr{H}_{\text{phys}}$ and thus eliminate the ghost states.

4. Derivation of the Johnson Sum Rule

In this section we will see that by canonical quantization of the Lagrangian (17) we get more information and in particular the Johnson sum rule (11). It was first observed by *Symanzik* [5] that (11) can be derived with the framework of the indefinite metric formulation. We start from the relation

$$C^{\mu} = V^{\mu} - \frac{1}{m}\partial^{\mu}C, \qquad [V^{\mu}, C] = 0, \tag{30}$$

which implies

$$[C^{\mu}C^{\nu}] = [V^{\mu}V^{\nu}] + \frac{1}{m^2}[\partial^{\mu}C\,\partial^{\nu}C].$$

Hence

$$[V^k V^0] = [C^k C^0] - \frac{1}{m^2}[\partial^k C, \partial^0 C],$$

where the first commutator vanishes. With (25) we obtain

$$[V^k(x)\,V^0(y)] = \frac{i}{Z_3 m_0^2}\partial^k \delta_3(x-y), \qquad x^0 = y^0$$

yielding Johnson's sum rule.

5. Finite Formulation of Field Equations

Due to the singular commutation relations of C_{μ} and ψ the local operator product $C_{\mu}(x)\psi(x)$ occurring in the Dirac equation (18) is not well-defined. In perturbation theory the divergence of this local product is compensated by assigning infinite values to M_0 and Z_2. However, then the field equation becomes meaningless in any mathematical sense. In order to give a well-defined meaning to the right hand side of (18) we replace it by the limit [9—13]

$$i\gamma^{\mu}\partial_{\mu}\psi = \lim_{\xi\to 0}\left\{g\frac{\gamma^{\mu}}{2}(C_{\mu}(x+\xi) + C_{\mu}(x-\xi))\,\psi(x) + M_0(\xi)\,\psi(x)\right\}. \tag{31}$$

$M_0(\xi)$ is a finite function of ξ, logarithmically divergent at $\xi = 0$. The value at $\xi = 0$ formally corresponds to the renormalization constant.

The meson field Eq. (19) we rewrite similarly

$$-\partial_{\nu}C^{\mu\nu} = \lim_{\xi\to 0}(Z_3(\xi)^{-1}\hat{j}^{\mu}(x\xi) - m_0^2(\xi)\,V^{\mu}(x)) \tag{32}$$

where $\hat{j}^{\mu}$ is a complicated expression with leading term

$$\hat{j}^{\mu}(x\xi) = \frac{1}{2}Z_2'(\xi)\,(\overline{\psi}(x+\xi)\,\gamma^{\mu}\psi(x-\xi) + \overline{\psi}(x-\xi)\,\gamma^{\mu}\psi(x+\xi)) + \cdots \tag{33}$$

For the complete expression $j^{\mu}(x\xi)$ one would like to have the following properties

(i) invariance under operator gauge transformations,

(ii) invariance under charge conjugation,

A form satisfying these requirements is*

$$\begin{aligned} j^\mu(x\xi) &= Z_2(\xi)\, \hat{J}^\mu(x\xi) + c(\xi)\, \xi^{\mu\nu}\, \hat{J}_\nu(x\xi) \\ &\quad + a_1(\xi)\, \xi_{\nu\kappa}\, \partial^\kappa A^{\mu\nu}(x) + a_2(\xi)\, \xi^{\mu\kappa}\, \partial^\nu A_{\kappa\nu} \\ &\quad + a_3(\xi)\, \xi^{\mu\nu} V_\nu(x) \end{aligned}$$

$$\hat{J}^\mu(x\xi) = \lim_{\xi\to 0} \frac{\hat{J}^\mu(x\xi\eta)}{y'(\xi\eta)}, \qquad \xi^{\mu\nu} = \frac{\xi^\mu \xi^\nu}{\xi^2} \tag{34}$$

$$\hat{J}^\mu(x\xi\eta) = \tfrac{1}{2}(\hat{Q}^\mu(x\xi\eta) + \hat{Q}^\mu(x_1 - \xi, -\eta))$$

$$\begin{aligned} \hat{Q}^\mu = {}&:\overline{\psi}(x+\xi)\, \gamma^\mu \exp\left\{ -i\frac{g}{2} \int_{x-\xi}^{x+\xi} \mathrm{d}y^\mu (C_\mu(y+\eta) \right. \\ &\left. + C_\mu(y-\eta)) \right\} \psi(x-\xi): \end{aligned}$$

: : denotes a generalized Wick product defined recursively by

$$O_1 \cdots O_n = : O_1 \cdots O_n : + \sum \langle O_{i_1} \cdots O_{i_\alpha} \rangle_0 : O_{i_{\alpha+1}} \cdots O_{i_n} :$$

for field operators C_μ or ψ at spacelike distances. The direction dependent terms were added in accordance with *Wilson*'s hypothesis [10] on the construction of local opeeator products. (34) has not yet been checked in perturbation theory. A different form for $\hat{j}^\mu$ (which is not invariant under operator gauge transformations) has been given by *R. Brandt* for the case of electrodynamics [11]. It is valid in perturbation theory and can presumably be extended to massive vector fields.

Concerning the renormalization functions $m_0^2(\xi)$ and $Z_V(\xi)$ we can make the following statement. Since the divergence (23) is a renormalized operator solution of the Klein-Gordon equation the commutator (25) must be a well-defined fourdimensional distribution and, of course, should not vanish. Hence

$$[C(x), C(y)] = -\frac{m^2}{\overline{m}^2}\, i\Delta(x-y) \tag{35}$$

with

$$\overline{m}^2 = \lim_{\xi\to 0} Z_V(\xi) m_0^2(\xi) \neq 0, \infty\,. \tag{36}$$

In the Proca-Wentzel formulation the meson field equation becomes

$$-\partial_\nu V^{\mu\nu} = \lim_{\xi\to 0} (Z_V(\xi)^{-1} j^\mu(x\xi) - m_0^2(\xi)\, V^\mu(x))\,, \tag{37}$$

$$j^\mu(x\xi) = \frac{1}{2} Z_2(\xi)\, (\overline{\varphi}(x+\xi)\gamma^\mu \varphi(x-\xi) + \overline{\varphi}(x-\xi)\gamma^\mu \varphi(x+\xi)) + \cdots \tag{38}$$

---

* In the original version of the manuscript it was incorrectly stated that the current (34) is conserved before the limit $\xi \to 0$. I am grateful to Dr. *Zumino* for pointing out this error to me.

The complete expression $j^\mu(x\xi)$ is again of the form (34) with C_μ, ψ replaced by V_μ, φ respectively and modified renormalization functions Z_2, Y.

6. Current-Field Relation

In this section we assume that the bare meson mass is infinite:

$$\lim_{\xi\to 0} m_0^2(\xi) = \infty . \tag{39}$$

Under this hypothesis a rigorous derivation of the current-field relation

$$j_\mu(x) \propto V_\mu(x)$$

of *Kroll*, *Lee* and *Zumino* [14] will be given. Equivalent to (37) is the relation

$$-\partial_\nu V^{\mu\nu} = g Z_V^{-1}(\xi) j^\mu(x\xi) - m_0^2(\xi) V^\mu(x) + O^\mu(x\xi)$$

with

$$\lim_{\xi\to 0} O^\mu(x\xi) = 0 .$$

Solving this equation for the current we obtain

$$j^\mu(x\xi) = \frac{Z_V(\xi) m_0^2(\xi)}{g} \left\{ V^\mu(x) - \frac{\partial_\nu V^{\mu\nu} + O^\mu(x\xi)}{m_0^2(\xi)} \right\} .$$

The second term vanishes in the limit and we get using (36)

$$j^\mu(x) = \lim_{\xi\to 0} j^\mu(x\xi) = \frac{\overline{m}^2}{g} V^\mu(x) \tag{40}$$

provided the bare meson mass is infinite.

The current-field identity (40) yields the surprising result that

$$[V^k(x), j^0(y)] = \frac{i}{g} \partial^k \delta_3(x-y) , \quad x^0 = y^0 . \tag{41}$$

Hence the assumption (13)

$$[V^k(x), j^0(y)] = 0 , \quad x^0 = y^0$$

is not correct if $m_0^2 = \infty$. Nevertheless Johnson's sum rule holds in general, as was shown in Section I, 4.

With the current-field identity also the current-current commutators follow

$$[j^0(x), j^0(y)] = [j^k(x), j^l(y)] = 0 ,$$

$$[j^k(x), j^0(y)] = i \frac{\overline{m}^2}{g^2} \partial^k \delta_3(x-y) .$$

Thus all the field and current commutators are determined by the canonical quantization rules.

II. Two Vector Fields Coupled to the Same Current

1. Two Massive Vector Fields

For this model the field equations in limit form are

$$-\partial_\nu V_k^{\mu\nu}(x) = \lim_{\substack{\xi\to 0\\ k=1,2}} (f_k(\xi) j^\mu(x\xi) - m_{k0}^2(\xi)\, V_k^\mu(x)) . \tag{42}$$

We will see that it is impossible to have two independent vector fields coupled to the same current provided the bare masses are infinite [6].

First we assume that the limit

$$j^\mu(x) = \lim_{\xi\to 0} j^\mu(x\xi) \tag{43}$$

exists. Then by an argument similar to the one used in Section I. 6 we obtain

$$V_1^\mu(x) \propto j^\mu(x) \propto V_2^\mu(x) \quad \textit{if} \quad \lim_{\xi\to 0} m_{k0}^2(\xi) = \infty .$$

Hence V_1^μ and V_2^μ are linearly dependent.

But we need not even assume the existence of the limit (43). The proportionality follows from the field equations alone and the hypothesis

$$\lim_{\xi\to 0} m_{0k}^2(\xi) = \infty .$$

For the proof we write the field equation in the equivalent form

$$-\partial_\nu V_k^{\mu\nu} = f_k j^\mu - m_{k0}^2\, V_k^\mu + O_k^\mu , \qquad \lim_{\xi\to 0} O_k^\mu(x\xi) = 0 .$$

Solving for V_k^μ we get

$$V_1^\mu = h_1(\xi) j^\mu(x\xi) + p_1^\mu(x\xi) , \tag{44}$$

$$V_2^\mu = h_2(\xi) j^\mu(x\xi) + p_2^\mu(x\xi) , \tag{45}$$

$$h_k(\xi) = \frac{f_k(\xi)}{m_{k0}^2(\xi)} , \qquad p_k^\mu(x\xi) = \frac{O_k^\mu(x\xi) + \partial_\nu V_k^{\mu\nu}(x)}{m_{k0}^2(\xi)} ,$$

$$\lim_{\xi\to 0} p_k^\mu = 0 .$$

Let ξ_n be a sequence of spacelike vectors with $\lim_{n\to\infty} \xi_n = 0$. Choose a subsequence ξ_n such that

$$\frac{h_1(\xi_n')}{h_2(\xi_n')}$$

has only one accumulation point. We make the assumption that

$$\lim_{n\to\infty} \frac{h_1(\xi_n')}{h_2(\xi_n')} = h \quad \text{is finite} . \tag{46}$$

This is no loss of generality. If (46) should be infinite one interchanges the subscripts 1 and 2. We multiply (45) by h_1/h_2 and take the limit

$$\lim_{n\to\infty} \frac{h_1(\xi'_n)}{h_2(\xi'_n)} V_2^\mu(x) = \lim_{n\to\infty} h_1(\xi'_n) j^\mu(x\xi'_n) = V_1^\mu(x).$$

Hence

$$V_1^\mu = h V_2^\mu.$$

Therefore, the two vector fields are linearly dependent if the bare masses are infinite.

2. Conventional Model of the Electromagnetic Field and a Vector-Meson Field Coupled to the Same Current

In general one should expect trouble if several vector fields are present and the number of linearly independent currents is smaller than the number of independent fields. Precisely this is the case in certain vector meson dominance models, but the essential difference is that in these models one vector field is massless, the electromagnetic field.

Let us treat the simplest case where two vector fields are coupled to the same current of which one is massless. The field equations are

$$-\partial_\nu F^{\mu\nu}(x) = \lim_{\xi\to 0} Z_A^{-1}(\xi)\,(e j^\mu(x\xi) - \partial^\mu \partial_\nu A^\nu(x)), \tag{47}$$

$$-\partial_\nu V^{\mu\nu}(x) = \lim_{\xi\to 0} \{Z_V^{-1}(\xi)\,(g j^\mu(x\xi) - m_0^2(\xi)\, V^\mu(x)), \tag{48}$$

$$F^{\mu\nu} = \partial^\mu A^\nu - \partial^\nu A^\mu.$$

This would correspond to the conventional way of coupling an electromagnetic field and a vector meson field to a current. But we will see that other formulations are possible which in fact work better than the conventional one. In the present model the field equations alone do not lead to a contradiction, but we will find an inconsistency if we use in addition the canonical commutation relations. For the purpose of quantization we choose the appropriate indefinite metric formulation for both fields. As Lagrange function we take

$$\mathscr{L} = \mathscr{L}_A + \mathscr{L}_C + \mathscr{L}_{\text{matt}} - (e_0 A_u^\mu + g_0 C_u^\mu) j_\mu. \tag{49}$$

$\mathscr{L}_A$ denotes the Lagrangian of the electromagnetic field [15]

$$\mathscr{L}_A = -\frac{1}{4} F_{u\mu\nu} F_u^{\mu\nu} - \frac{1}{2} Z_3^{-1} (\partial_\mu A_u^\mu)^2,$$

$$A^\mu = Z_3^{-\frac{1}{2}} A_u^\mu, \qquad e = Z_3^{\frac{1}{2}} e_0$$

The remaining notations are as in Section I. 2. As field equations we obtain (47) and

$$-\partial_\nu C^{\mu\nu} = \lim_{\xi\to 0}(g Z_C^{-1}(\xi) j^\mu(x\xi) - m_0^2(\xi)\, V^\mu(x))\,, \tag{50}$$

$$\Box\, \partial_\mu A^\mu = 0\,, \qquad (\Box + m^2)\, \partial_\mu C^\mu = 0\,. \tag{51}$$

For the Proca field (27) the field equation (48) follows. For infinite bare meson mass

$$\lim_{\xi\to 0} m_0^2(\xi) = \infty \tag{52}$$

we obtain again the current-field relation

$$\lim_{\xi\to 0} j^\mu(x\xi) = \frac{\overline{m}^2}{g}\, V^\mu(x) \tag{53}$$

yielding

$$-Z_3\,\partial_\nu F^{\mu\nu} + \partial^\mu \partial_\nu A^\nu = \frac{e}{g}\,\overline{m}^2\, V^\mu\,, \qquad Z_3^{-1} = \lim_{\xi\to 0} Z_3^{-1}(\xi)\,. \tag{54}$$

Using canonical quantization we find

$$[V^k(x)\, V^0(y)] = \frac{1}{\overline{m}^2}\, i\,\partial^k \delta_3(x-y)\,, \qquad x^0 = y^0\,.$$

On the other hand the canonical rules imply that V^0 commutes with $\partial_\mu A^\mu$ and $\partial_0 F^{k0}$ which is in contradiction to the field equation (54).

3. Kroll-Lee-Zumino Model [14]

This is another model of an electromagnetic field and a vector meson field coupled to the same current. Compared to the conventional model the main advantage is that the current-field relation already holds for finite bare mass. The Lagrangian of this model is

$$\mathscr{L}' = \mathscr{L}_A + \mathscr{L}_1 + \mathscr{L}_2 + \mathscr{L}_{\mathrm{matt}} - (e_0 A_{u\mu} + g_0 C_{u\mu}) j^\mu\,,$$

$$\mathscr{L}_1 = -\frac{1}{4}\left(C_{u\mu\nu} + \frac{e_0}{g_0} A_{u\mu\nu}\right)\left(C_u^{\mu\nu} + \frac{e_0}{g_0} A_u^{\mu\nu}\right), \tag{55}$$

$$\mathscr{L}_2 = \frac{1}{2}\, m_0^2 C_{u}{}^{\mu} C^\mu + \frac{1}{2}\,\frac{m_0^2}{m^2}\,(\partial_\mu C_u^\mu)^2\,;$$

(55) is related to the Lagrangian (49) of the preceding Section by

$$\mathscr{L}' = \mathscr{L} - \frac{1}{2}\,\frac{e_0}{g_0}\, C_{u\mu\nu} F_u^{\mu\nu} - \frac{1}{4}\,\frac{e_0^2}{g_0^2}\, F_u^{\mu\nu} F_{u\mu\nu}\,.$$

Essential is that in addition we have a mixing term of the two fields, the last term simply changes the normalization of the electromagnetic field. The field equations in limit form are

$$-\partial_\nu F^{\mu\nu} = \lim_{\xi\to 0}\left\{ eZ_3^{-1}(\xi) j^\mu(x\xi) - Z_3^{-1}(\xi)\,\partial^\mu\partial_\nu A^\nu - \frac{e^2}{g^2}\frac{Z_V(\xi)}{Z_3(\xi)}\,\partial_\nu F^{\mu\nu} + \frac{e}{g}\frac{Z_V(\xi)}{Z_3(\xi)}\,\partial_\nu C^{\mu\nu}\right\}, \tag{56}$$

$$-\partial_\nu C^{\mu\nu} = \lim_{\xi\to 0}\left\{ Z_V(\xi)^{-1} g j^\mu(x\xi) - m_0^2(\xi)\,V^\mu(x) + \frac{e}{g}\,\partial_\nu F^{\mu\nu}\right\}, \tag{57}$$

$$\Box\,\partial_\mu A^\mu = 0\,, \qquad (\Box + m^2)\,\partial_\mu C^\mu = 0\,. \tag{58}$$

Using (58) and canonical quantization we obtain again (35) and (36).

For the purpose of eliminating $\partial_\nu C^{\mu\nu}$ in (56) we write (56) and (57) in the equivalent form

$$-\partial_\nu F^{\mu\nu} = \{\cdots\} + w^\mu\,, \tag{59}$$

$$-\partial_\nu C^{\mu\nu} = \{\cdots\} + v^\mu \tag{60}$$

with

$$\lim_{\xi\to 0} w^\mu(x\xi) = 0\,, \qquad \lim_{\xi\to 0} v^\mu(x\xi) = 0\,.$$

Multiplying (60) by $-\dfrac{eZ_V}{gZ_3}$ and inserting the result into (59) one gets

$$-Z_3\partial_\nu F^{\mu\nu} = \frac{eZ_V m_0^2}{g}\,V^\mu - \frac{e}{g}\,Z_V v^\mu + Z_3 w^\mu\,. \tag{61}$$

If

$$\lim_{\xi\to 0} Z_3(\xi) = 0$$

one gets the trivial relation

$$\frac{e\overline{m}^2}{g}\,V^\mu = \partial^\mu\partial_\nu A^\nu\,,$$

which implies that V^μ is a free field of mass zero. Hence the renormalization constant of the electromagnetic field

$$Z_3^{-1} = \lim_{\xi\to 0} Z_3^{-1}(\xi) \neq \infty \tag{62}$$

must be finite [16]. From (61) we obtain in the limit $\xi\to 0$

$$-Z_3\partial_\nu F^{\mu\nu} = \frac{e\overline{m}^2}{g}\,V^\mu - \partial^\mu\partial_\nu A^\nu\,, \tag{63}$$

which is identical to the field equation (54) of the conventional model. Eq. (63) shows that the electromagnetic current is

$$j^\mu = \frac{\overline{m}^2}{g} V^\mu \tag{64}$$

no matter whether the bare meson mass is finite or not. No contradiction could be found if (63) is combined with the canonical quantization rules based on (55).

Footnotes and References

1. *Proca, J.*: J. Phys. Radium **7**, 347 (1936). — *Wentzel, G.*: Quantum theory of fields. New York: Interscience Publishers Inc. 1949.
2. *Johnson, K.*: Nucl. Phys. **25**, 435 (1961); **31**, 464 (1962).
3. *Ogievetskii, O. V. I.*, and *I. V. Polubarinov*: J. Exptl. Theoret. Phys. (U.S.S.R.) **41**, 247 (1961); — JETP **14**, 179 (1962).
4. *Feldman, G.*, and *P. T. Matthews*: Phys. Rev. **130**, 1633 (1963).
5. *Symanzik, K.*: unpublished.
6. *Zimmermann, W.*: Commun. Math. Phys. **8**, 66 (1968).
7. The commutator $[V^\mu, C] = 0$ follows from (22) and the canonical rules applied to $\mathscr{L}'$ (Eq. (17)). The commutator $[\varphi, C] = 0$ follows from (22) and the canonical rules applied to $\mathscr{L} + \mathscr{L}_C$ with $\mathscr{L}$ given by (1).
8. Eq. (28) involves divergencies. For a more careful definition in limit form see Ref. [6], Appendix.
9. *Valatin, J.*: Proc. Roy. Soc. London A, **225**, 535 and **226**, 254 (1954).
10. *Wilson, K. G.*: Cornell Univ. Rep. 1964.
11. *Brandt, R. J.*: Ann. Phys. **44**, 221 (1967), Technical Report No. 673, University of Maryland and Annals of Physics, to be published.
12. *Zimmermann, W.*: Commun. Math. Phys. **6**, 161 (1967).
13. $\lim_{\xi \to 0}$ will denote the spacelike limit with $\xi^2 < 0$ and ξ^μ/ξ^2 bounded.
14. *Kroll, N. M.*, *T. D. Lee*, and *B. Zumino*: Phys. Rev. **157**, 1376 (1967). — *Zumino, B.*: Proceedings of the Heidelberg International Conference on Elementary Particles, p. 465 (1967).
15. *Gupta, S. N.*: Proc. Phys. Soc. London A **64**, 426 (1951). — *Källén, G.*: Handbuch der Physik, Vol. V, Part 1. Berlin-Göttingen-Heidelberg: Springer 1958.
16. In perturbation theory this was proved by *Lee* and *Zumino* (to be published in Nuovo Cimento).

Prof. Dr. *Wolfhart Zimmermann*
Courant Institute of Mathematical Sciences
251 Mercer Street
New York, N. Y. 10012/USA